COMMON FOSSIL PLANTS OF WESTERN NORTH AMERICA

William D.

Illust
Naomi E. Hebbert
Paul H. Smith

Brigham Young University Press
Provo, Utah

Library of Congress Cataloging in Publication Data

Tidwell, William D, 1932-
   Guide to the common fossil plants of western North
America.

   Includes index.
   1.   Paleobotany—North America—Pictorial works.
I.   Hebbert, Naomi.   II.   Smith, Paul, 1952-
III.   Title.
QE935.T5   561'.19'7   75-4640
ISBN 0-8425-1298-5
ISBN 0-8425-1301-9 pbk.

Brigham Young University Press, Provo, Utah 84602
©1975 by Brigham Young University Press. All rights
   reserved
Printed in the United States of America
75 10M 4479

Contents

Acknowledgments   5
Preface   6
Introduction   7
Purpose   8
Historical Review   9
Plant Kingdom   11
Types of Preservation and Methods of Study   27
Collecting and Processing Specimens   31
Sequence of Plant Life of the Past   35
Algae   49
Lycopodophyta   51
Sphenophyta   55
Filicophyta (Ferns)   59
Fernlike Foliage   68
Seeds and Microsporangiate Structures   71
Cycadophyta (Cycads)   75
Ginkgophyta (Ginkgos)   81
Coniferophyta   82
Anthophyta (Angiosperms)   95
Petrifaction   117
Petrified Filicophyta (Ferns)   124
Petrified Cycadophyta (Cycads)   127
Petrified Coniferophyta (Conifers)   129
Petrified Anthophyta (Angiosperms)   137
Glossary   150
Selected Readings   160
Index   174

Illustrations

Index Map to Fossil-plant Localities   22
Table 1—Plant Kingdom   12
Table 2—Plant Classification   21
Table 3—Geologic Time Scale   25
Table 4—Some Collecting Sites   31
Chart 1—Distribution of Genera in Carboniferous
    Floras   36
Chart 2—Distribution of Genera in Some Mesozoic
    Mesozoic Floras   40
Chart 3—Distribution of Genera in Upper
    Cretaceous and Tertiary Floras   following 48
Chart 4—Key to Some Cycadeoidean Foliage   80
Plates 1-48   following 116
Outline Key   162
Leaf Characters (Figures 326-331)   192

Acknowledgments

The author extends his appreciation to Dr. J. Keith Rigby of the Department of Geology, Brigham Young University, for his constant encouragement and assistance in reviewing the manuscript. Others I wish to thank for reviewing the manuscript at various stages are: Jay Shuler of McClellanville, South Carolina; Parker Chipman and Marian Whitehead of Salt Lake City, Utah; Frank and Leona Lemons of Moab, Utah; Thomas and Beth Hopkins of Hanksville, Utah; and Thomas Leslie of Provo, Utah. Thanks also go to Dr. Sidney Ash of Weber State College, Ogden, Utah (for the use of his key on cycadeoidean foliage); and Stuart Heimdal of Provo, Utah (for the use of his painting of Florissant, Colorado, during the Oligocene). Special thanks are due those who have donated specimens and in other ways contributed to this book.

Preface

The major purpose of this guide is to introduce the reader to a small portion of the fascinating world of nature through the study of fossil plants. The world of fossil plants is open to both professional and nonprofessional collectors, but the former often acquire a language so technical that their knowledge of plants is generally communicated to only a few. Technical terminology has been held to a minimum in this guide. However, some terms are essential in identifying, describing, and understanding fossil plants. These are defined and illustrated throughout the book as well as in the Glossary.

This guide has been designed to aid in identifying fossil plants that collectors will likely encounter. Each plant group is introduced with its basic terminology and a summary of its characteristics. The ecological conditions under which some of the extinct plants may have grown are discussed. A narrative style is used in the guide where the plant forms are closely associated or related, and a more descriptive style is used where the fossilized plant remains are distantly related to other plant types.

Drawings will aid the reader in understanding the descriptive portion of the text, and photographs will demonstrate to the collector how the actual specimens may look. Since few specimens will be absolutely perfect, the drawings are more idealized than the photographs.

Introduction

Probably all of us make our most frequent contact with nature through plants, and possibly no
part of our environment appeals to us more.
Certainly, the more we know about plants, the
more interesting they become. Paleobotany, the
study of fossil plants, is an exciting science that
helps us understand plant life of the past. With
only a little knowledge of botany, an individual
can understand the relationship of fossil plants to
the plants of today. A knowledge of botany also
contributes to our enjoyment of life.

Natural curiosity is an additional incentive for
studying fossil plants. Our wish to understand
why things are as we find them and how they
relate to each other probably has contributed
more to the growth of science than all other
influences combined. And a study of paleobotany will help to satisfy the natural curiosity
about fossilized plants.

This book is about plants from our ancient
past and their preservation in the crust of the
earth. Collecting is like opening Christmas pack-
ages. For a collector, the thrill of breaking
or splitting a rock and finding fossilized plant
remains that have been extinct for millions of
years is very rewarding. Who knows—the next
blow may uncover a fossil plant never before
known to man, or one that will aid further in
deciphering the earth's history. The humblest
collector has as good a chance of finding a beautiful, unique specimen as does the most learned
professional. It is hoped that this book will help
in the collector's understanding of the plant life
of the past and present.

7

Purpose

The purposes of this guide are to (1) discuss the various types of fossil plant preservation, (2) give an overview of the plant kingdom, (3) explain how fossil plants are named, (4) discuss the development of western floras, (5) describe the more common elements of the fossil floras of this region, and (6) include some of the better-known fossil-plant collecting sites. By using the guide, the collector will become better acquainted with the terms, information, and features commonly used in the study of fossil plants and in botany in general.

An important purpose for the guide is to help the collector recognize the scientific importance of his fossil plants. Good collections or unique fossil materials should be studied by professional paleobotanists to ascertain their significance in the history of plant life. Many important events in this history have been reconstructed using a specimen or collection donated by a collector for scientific study. For example, the first known specimens of *Osmundacaulis* (p. 124) from the state of Utah were donated by an amateur collector. The specimens were a new species *(Osmundacaulis wadei)* and were subsequently named for the collector.

Historical Review

The study of fossil plants has a long and varied history. They were recognized as early as the 6th century B.C., when laurel-leaf compressions were noted by the Greek Xenophanes. Fossil plant remains were mentioned by Leonardo Da Vinci in the 14th and 15th centuries and by Martin Luther in the 16th. With the aid of the first microscope, Robert Hooke described cell structures in fossil woods in 1664. Man's use of coal for heating and industry led to the early recognition of plant remains associated with these coals in Europe and, eventually, in the United States. In fact, during much of the 18th century, fossil plant studies were utilized to determine whether the biblical flood had occurred in the month of May or in the fall. The young cones Johann Scheuchzer illustrated in his book, *Herbarium Diluvianum* (1709), were presented as proof that the flood had occurred in May. However, Parsons (1757) thought that the abundance of "ripe" fossil fruits indicated that the flood had taken place in autumn.

A method for grinding thin sections developed by William Nicols (1768-1851) allowed for the first study of the anatomy of fossil plants.

Many scientific studies were reported on fossil plants during the early part of the 19th century. In France, Adolphe Brongniart, known as the Father of Modern Paleobotany, published a series of early works (1820-1838)—mostly on Carboniferous plants. Other paleobotanists, such as Charles H. Sternberg in Austria and E. H. von Schlotheim in Germany, were working in Europe at approximately this same time.

9

One of the first American paleobotanists, Leo Lesquereux, was originally a Swiss bryologist who immigrated to the United States. He published the *Coal Flora of Pennsylvania* in the years 1879-83. He did further studies on Carboniferous floras in other eastern states and on some Cretaceous and Tertiary floras of the western United States. Other early investigators of fossil plants include David White, who published a monograph on Missouri coal plants; G. R. Wieland, who researched fossil cycads of the Black Hills, South Dakota; and Sir William Dawson, an early Canadian paleobotanist. Today there are numerous researchers in the field, and the interest in the scientific study of fossil plants continues to expand.

Plant Kingdom

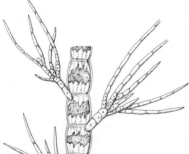

Figure 1. The freshwater green alga *Draparnaldia*

Figure 2. *Laminaria,* a brown alga

The principal subdivisions of the plant kingdom are illustrated in Table 1. Algae and fungi, although abundant in modern floras, are not commonly preserved as fossils. Bryophytes (mosses and liverworts) are also rarely fossilized. Most common fossil plant remains are portions of vascular plants.

Algae

Algae, the simplest green plants, have been grouped with the fungi into the Thallophytes, or plants without true roots, stems, leaves, and conductive systems. They occur as single cells, groups of cells (Fig. 1), or masses of cells in which they may be united variously to form filaments or spherical aggregates. Algae generally grow in water and therefore are adapted to a water environment. Freshwater algae grow in streams, lakes, and ponds (where they are often called pond scum). They also grow on stream banks and tree bark. Brown and red algae, commonly known as seaweed, grow attached to rocks in shallow marine or brackish water along seacoasts.

Algae vary in size. Some are so small as to be observed only under a microscope. The larger forms of seaweed have a relatively high degree of body complexity (Fig. 2) and attain lengths of a hundred or more feet. These are attached at their bases by strong holdfasts resembling roots.

All algae contain chlorophyll and are able to manufacture their own food. However, not all are green. In many, the chlorophyll is masked

11

with other coloring materials or pigments. The divisions of algae, therefore, are named for the general color of their plant bodies. Thus, the division Cyanophyta are the blue green algae (Greek *kyanos:* "dark blue enamel"), and the Rhodophyta are the reds (Greek *rhodon:* "rose").

Fungi

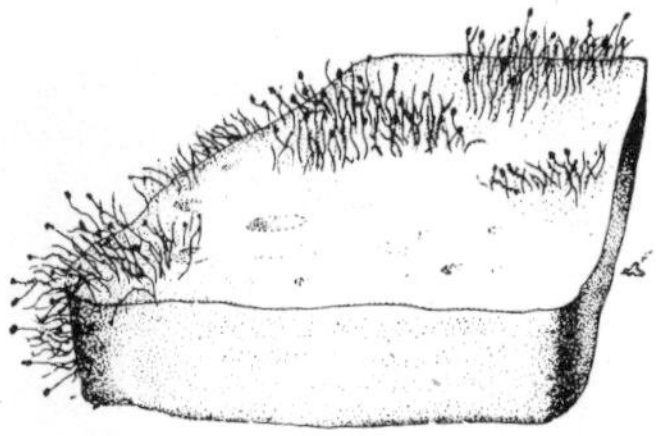

Figure 3. *Rhizopus* (bread mold)

Fungi basically differ from other members of the plant kingdom in not containing chlorophyll and, consequently, in not producing their own food. They obtain their sustenance from other organisms in two ways: (1) directly from living plants or animals as parasites, and (2) from organic substances produced by other plants or animals. The second group, called saprophytes, contains the greater number of fungi. An example is the common bread mold (Fig. 3). Saprophytes serve a useful purpose as scavengers, helping to dispose of dead plants and animals.

Table 1—Plant Kingdom

"Algae"

      Cyanophyta—blue green algae—Cryptozoans*
      Chlorophyta—green algae
      Chrysophyta—golden brown algae—Diatoms
      Phaeophyta—brown algae
      Rhodophyta—red algae—Corallinaceae (calcareous)

Fungi

Bryophyta—mosses and liverworts
Ferns and Fern Allies
    Psilophyta—*Psilotum*
    Lycopodophyta—club mosses—*Lepidodendron**
    Sphenophyta—*Equisetum* ("joint grass" or horsetails)—*Calamites**
    Filicophyta—*ferns*

"Gymnosperms"

    Pteridospermophyta—seed ferns*
    Cycadophyta
        Cycadales—living cycads—*Zamia*
        Cycadeoidales—fossil cycads—*Cycadeoidea**
    Ginkgophyta—*Ginkgo*—(Maidenhair tree)

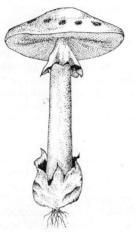

Figure 4. Mushroom (*Amanita*)

Figure 5. The liverwort *Marchantia*

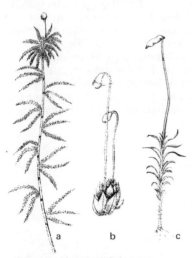

Figure 6. Mosses. a. *Sphagnum* (bog moss), b. *Funaria,* c. *Polytrichum.* (after *Plants and Man,* Rushforth and Tidwell)

Coniferophyta
 Cordaitales—*Cordaites**
 Coniferales—pines and firs
 Gnetophyta—*Ephedra*

Anthophyta—Angiosperms—flowering plants
 Dicotyledonae—oaks and maples
 Monocotyledonae—corn and palms

*extinct

Fungi may include bacteria (the simplest fungal form), slime molds, algal fungi (including the water molds that infect fish), sac fungi (yeasts and various blights), and the higher fungi (mushrooms [Fig. 4], puffballs, rusts, and smuts of cereal crops).

Bryophyta

Bryophyta are considered to be more advanced plants than algae and fungi. Their plant body, however, is similar in that it is a thallus. The bryophyta is composed of two major groups: (1) the liverworts or Hepaticae and (2) the mosses. The mosses (Musci) are more complex than the liverworts and have stemlike and leaflike structures similar in appearance to those of higher plants.

Liverworts are found in a variety of environments. Some float, but most grow in damp places on the ground. Their body is a thin, flat, lobed, leaflike thallus, not erect, but in close contact with the soil or material upon which it is growing (Fig. 5). These plants seem only slightly modified for life on land. Their name, Hepaticae, originates from their fancied resemblance to the lobes of the human liver, and because of that resemblance they were once considered to be medicinally useful for liver ailments.

Mosses are abundant and familiar plants (Fig. 6). Small, with thin, simple "leaves" spirally arranged about an often upright central axis, these leafy-stemmed plants are anchored in the soil by delicate rhizoids. In bogs composed of sphagnum moss, the lower moss layers do not completely decay; instead, they build up with other plant debris into a coallike substance called peat.

13

Figure 7. *Lycopodium*

Figure 8. *Selaginella* (club moss)

Figure 9. *Isoetes* (quillwort)

Ferns and Fern Allies

Ferns and fern allies constitute a large assemblage of plants. Like the lower forms of plant life (algae, fungi, and bryophytes), they reproduce by spores rather than seeds. However, like the seed plants, ferns and their allies possess water-conducting and food-conducting tissues, which the lower plants lack. Geologically, they appeared earlier than the seed plants and were fairly abundant during the Carboniferous period. Presently three groups of the ferns and their allies are rather common in North America: Lycopodophyta or club mosses, Sphenophyta or joint-grass, and Filicophyta or true ferns. The spore plants of Sphenophyta and Lycopodophyta differ from true ferns in external appearance and structure and are not readily recognized as fern allies by the casual observer.

The club mosses or Lycopodophyta resemble mosses with coarse leaves, as the name suggests. Their leaves are small, and they grow in clusters or mats on the forest floor of temperate and tropical regions. Club mosses are slender, branching plants with creeping, trailing, or erect stems completely clothed with small leaves. Known by names such as "ground pine" and "running cypress," they have been used for Christmas decorations.

There are three orders of living plants in the Lycopodophyta. The first, Lycopodiales (genus *Lycopodium*) (Fig. 7), is homosporous; that is, it produces only one size of spore. The other two, Selaginellales (genus *Selaginella*) (Fig. 8) and Isoetales (genus *Isoetes*) (Fig. 9), produce two sizes, a megaspore (female) and a microspore (male), and are called heterosporous. These two kinds of spores are formed in different spore-producing structures (sporangia) on the same plant.

14

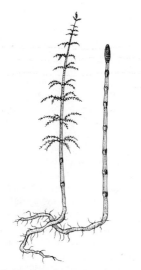

Figure 10. *Equisetum* (horsetail)

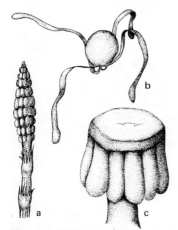

Figure 11. *Equisetum* (horsetail). a. Cone or strobilus, b. Spore with elators, c. Close-up of sporangiophore

During the Carboniferous period, lycopods formed a chief part of the forests, including large trees of various types, such as lepidodendrons and sigillarians (Fig. 34). Only the smaller forms have lived to the present time.

In swamps, on embankments, and along streambanks can be seen slender columnar plants with conspicuously jointed stems having whorled scale leaves and spore-bearing cones on the tips of their upright branches. These are common horsetails, scouring rushes, or joint grasses, which are known scientifically as *Equisetum.* (Fig. 10). *Equisetum* has an underground stem or rhizome that serves to store food for dormant periods, when the exposed aerial plant parts are destroyed by drought or freezing. From this rhizome grow the annual green shoots, unbranched in some species and multibranched in others. The cones are composed of a series of whorled shield-shaped sporangiophores (Fig. 11a, 11c). Each sporangiophore bears several sporangia shaped like fingers on a glove. Spores of *Equisetum* have four appendages or elators that expand when wet and contract when dry (Fig. 11b). These appear to aid in spore distribution. Because the spores are of the same size, the plant is homosporous.

Equisetum is represented in modern floras by 25 species; but during the Carboniferous the species of Sphenophyta were more numerous, primarily under the genera *Sphenophyllum* and *Calamites.* Some species of the latter genus were large trees and constituted a conspicuous portion of the Carboniferous forest.

15

Figure 12. Fern

Figure 13. Tree fern

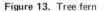

Figure 14. Fossil cycadeoid
Williamsonia

The Filicophyta or true ferns are well-known plants, and the common forms are easily recognized. They are noted for their large leaves. True ferns are generally land plants and are therefore unlike mosses and liverworts in their structure and manner of life (Fig. 12). They grow in considerable numbers on the floors of forests where the light is diminished by the higher tree foliage and the soil is moist and fertile with much decaying humus. Ferns are a striking and characteristic part of tropical vegetation, forming masses of smaller plants as well as large tree ferns. Tree ferns have cylindrical trunks 35 to 45 feet high, with a crown of leaves at the stem tip (Fig. 13). The trunk is encased by persistent leaf bases that give it a rough appearance.

The stem of most ferns is an underground rhizome from which young leaves spring and uncoil in a peculiar circinate or fiddlehead manner (p. 60). Cell structure and tissues of leaves, stems, and roots of the ferns and their allies are similar to those in the higher seed plants.

Members of the Filicophyta, Sphenophyta, and Lycopodophyta were associated during the Carboniferous Period and were very important in the formation of coal. Swamps covered immense areas, supporting a luxuriant vegetation. Plant remains accumulated in thick deposits in waters of these swamps, and when the land sank, sediments were deposited on this decaying vegetation. The weight of the overlying sediments that were uplifted to form mountains caused tremendous pressure and, as a result, heat. This heat drove out the hydrogen and oxygen from these organic remains, leaving only the carbon, which solidified into coal. The type of coal depends upon the amount of change wrought upon the vegetation. Hard or anthracite coal, which has changed the most, consists of 85 percent or more carbon, whereas bituminous or soft coal contains only 50 to 75 percent. When coal is burned today, carbon dioxide returns to the atmosphere that it came from millions of years before.

Figure 15. Seed fern

Figure 16. Living cycad *Cycas*

Figure 17. Living cycad

"Gymnosperms"

Seed plants make up the most conspicuous portion of the vegetation on earth today. They are the major sources of food and lumber and are the most familiar plants. The two important groups of seed plants are the gymnosperms and the angiosperms or flowering plants. The name *gymnosperm*, from the Greek *gymnos*, "naked," and *sperm*, "seed," suggests the group's most characteristic trait: seeds of members of this informal group are borne naked upon the surface of the sporophylls (which constitute the cone) rather than within the ovary, as are the seeds of angiosperms.

Plant groups included in the gymnosperms are seed ferns, cycads, ginkgos, conifers, and Gnetophyta, each of which will be considered separately.

Pteridospermophyta

This division constitutes the seed ferns, a group of extinct plants with leaves that closely resemble those of ferns but which bore seeds (Fig. 15). Seed ferns were an important part of Carboniferous floras.

Cycadophyta

Living cycads are subtropical to tropical forms that have changed little over the last 200 million years (Fig. 16). Their plant body consists of a relatively short, thick, columnar stem covered with numerous leaf bases topped with a crown of large fernlike leaves (Fig. 17). Though cycads are similar in appearance to palms, the two forms are not closely related. Male and female cones of cycads are borne externally on separate plants near the center of the crown of leaves.

Figure 18. *Ginkgo* with male cone

Ginkgophyta

Ginkgo (maidenhair tree) is the only surviving member of this group of ancient trees. The group is known in the geologic record primarily from fossilized leaves. The leaves of the living plants are rather leathery and are lobed and veined like those of *Adiantum* or maidenhair fern; hence the common name for *Ginkgo* (Fig. 18). Male and female cones are borne on short shoots on separate trees.

Coniferophyta

Conifers today form extensive forests in the temperate zones. Their leaves may be scaly, broad, or needlelike. Red cedar and arborvitae leaves are scalelike, whereas pine, spruce, and hemlock leaves are needlelike. The trunks of conifers are large, woody, and many-branched (Fig. 19). Conifer wood lacks the water-conducting tubes or vessels common in the wood of angiosperms. The wood cells or tracheids in conifer wood perform a double function, conducting water and supporting the tree. Many conifers, such as pines, have resin tubes or ducts extending throughout the plant, even in the leaves; others, like *Sequoia* (redwood), may have only resin cells. Seeds are produced in cones. Some seeds, such as those in pinyon pine, are usually large. Pinyon pine seeds or pine nuts furnished much food for the Indians of California and the lower Great Basin.

Figure 19. Living conifers

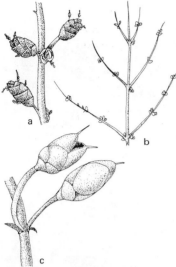

Figure 20. *Ephedra* (Mormon tea). a. Male cone, b. *Ephedra* plant, c. Female cone

Figure 21. *Welwitschia*

Gnetophyta

Gnetophyta consists of three orders with one genus in each. Ephedrales *(Ephedra)* occurs in the desert regions around the world. The stems of this plant are photosynthetic, and the scalelike leaves are whorled like those of *Equisetum*. However, *Ephedra*, commonly called Mormon tea, produces naked seeds in cones and is, therefore, considered a gymnosperm—as are the other two orders (Fig. 20). Welwitschiales *(Welwitschia)* grows in an extremely dry desert in a small part of South Africa. They are odd-looking plants bearing two opposite leaves around a wooden, bowl-shaped stem (Fig. 21). These leaves become battered and tattered with age. Gnetales *(Gnetum)*, which grows in the luxuriant tropics, is a viny plant with leaves very similar to those of angiosperms (Fig. 22). But again, they produce seeds in cones as do the other gymnosperms.

Anthophyta (angiosperms or flowering plants)

Angiosperms include the common herbaceous and woody-stemmed plants such as grasses, clovers, elms, oaks (Fig. 25), and maples. Unlike other forms of seed plants, this large and important group contains a striking diversity of vegetative and reproductive structures.

Diversity allows the angiosperms to grow in a variety of land and water habitats, from polar regions to tropics and from oceans to high mountain peaks. Brightly colored floral parts are found in most members of this group and they attract pollinators such as bees, moths, or bats for pollen transfer. The seeds that subsequently develop are normally protected within the enclosed ovary. This ovary is often secured within a fruit which, in turn, aids in the distribution of the seeds.

Figure 22. *Gnetum.* a. Leaf and fruits, b. Female cone

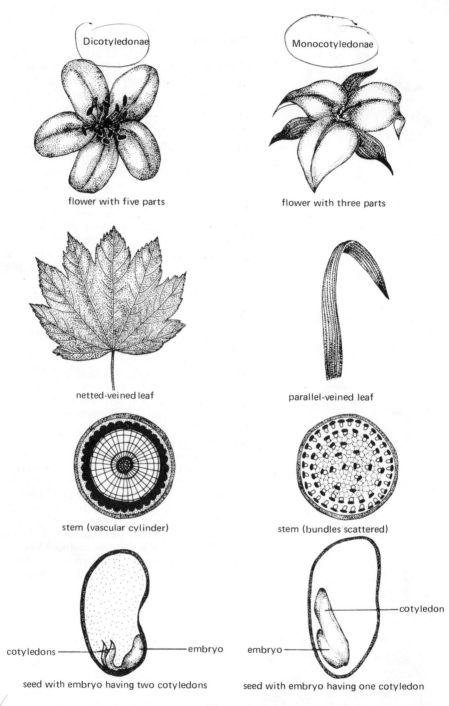

Dicotyledonae

Monocotyledonae

flower with five parts

flower with three parts

netted-veined leaf

parallel-veined leaf

stem (vascular cylinder)

stem (bundles scattered)

cotyledons — embryo

embryo

cotyledon

embryo

seed with embryo having two cotyledons

seed with embryo having one cotyledon

Figure 23. Dicot-monocot comparisons

The angiosperms are divided into two classes: the Monocotyledonae and the Dicotyledonae (Fig. 23).

Monocotyledonae

Monocots usually have their floral parts in groups of three (e.g., three stamens). Their vascular bundles are scattered throughout the stem; their leaves are generally parallel veined; and they possess only one cotyledon or seed leaf in their embryo. Their stems are predominately herbaceous and rarely have secondary growth. The monocots include palms (Fig. 24), lilies, and grasses. Grasses produce the bulk of the food consumed by man and animals.

Figure 24. Living palm

Dicotyledonae

Dicot floral parts occur in groups of four or five. Their bundles are arranged in a circle around the pith of the stem. They have a large number of species with secondary growth. Generally their leaves are net veined, and their embryos contain two cotyledons. In this group belong the broad-leafed trees (Fig. 25) and shrubs and a large number of herbaceous forms.

Plant Classification

In plant classification (Table 2), related plants are grouped by a system of names (taxonomy). The use of common names is not always practical, as they vary from region to region, and many plants do not have them. Scientific names are used because they are the same in any language. They are derived from Latin or Greek words and consist of two parts, like a person's name. This is known as binomial nomenclature. The first or generic portion of the name denotes plants with close affinities. The second part, the species name, is specific to the designated plant. For example, the wild tiger lily is known scientifically as *Lilium columbianum.* There are many species of lilies under the genus *Lilium,* but only one species is named *columbianum.* This is its binomial. The person who originally proposed

Figure 25. Living *Quercus* (oak)

Table 2—Plant Classification

	Kingdom—Plant
	Division—Anthophyta
	Class—Monocotyledonae
	Order—Liliales
Smith Family Organization	Family—Liliaceae
Smith Family (group of closely related individuals)	Genus—Lilium
Smith, Paul (a certain individual)	Species—*Lilium columbianum* Hanson (wild tiger lily)

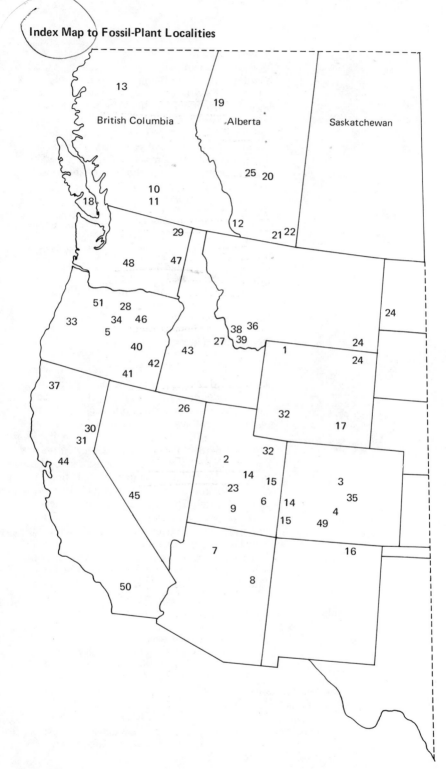

Index Map to Fossil-Plant Localities

British Columbia

Alberta

Saskatchewan

Paleozoic
1. Beartooth Butte flora (Dev.)
2. Manning Canyon Shale flora (Penn.)
3. Mosquito Range flora (Penn.)
4. Salida Canyon flora (Penn.)
5. Spotted Ridge flora (Penn.)
6. Honaker Trail flora (Penn.)
7. Hermit Shale flora (Perm.)

Mesozoic
8. Petrified Forest N.P. flora (Tri.)
9. Morrison Formation flora (Jur.)
10. Spence Bridge flora (L. Cret.)
11. Jackass Mountain flora (L. Cret.)
12. Kootenay Formation flora (L. Cret.)
13. Smithers flora (L. Cret.)
14. Cedar Mountain flora (L. Cret.)
15. Dakota Sandstone flora (M. Cret.)
16. Raton Formation flora (U. Cret.)
17. Medicine Bow flora (U. Cret.)
18. Nanaimo Group flora (U. Cret.)
19. Dunvegan flora (U. Cret.)
20. Edmonton flora (U. Cret.)
21. Milk River flora (U. Cret.)
22. Frenchman flora (U. Cret.)
23. Blackhawk flora (U. Cret.)

Cenozoic
24. Fort Union flora (Paleoc.)
25. Paskapoo Formation flora (Paleoc.)
26. Copper Basin flora (Eoc.)
27. Salmon flora (Eoc.)
28. Clarno flora (Eoc.)
29. Republic flora (Eoc.)
30. LaPorte flora (Eoc.)
31. Chalk Bluffs flora (Eoc.)
32. Green River flora (Eoc.)
33. Comstock flora (Eoc.)
34. Bridge Creek flora (Olig.)
35. Florissant flora (Olig.)
36. Ruby flora (Olig.)
37. Weaverville flora (Olig.)
38. York Ranch flora (Olig.-Mio.)
39. Beaverhead flora (Olig.-Mio.)
40. Stinking Water flora (Mio.)
41. Trout Creek flora (Mio.)
42. Sucker Creek flora (Mio.)
43. Thorn Creek flora (Mio.)
44. San Pablo flora (Mio.)
45. Fingerrock flora (Mio.)
46. Mascall flora (Mio.)
47. Latah flora (Mio.)
48. Ellensberg flora (Mio.-Plio.)
49. Creede flora (Plio.)
50. Mount Eden flora (Plio.)
51. Deschutes flora (Plio.)

the species name may be added at the end; hence, *Lilium columbianum* Hanson.

Similar genera are grouped together into families, families into orders, orders into classes, and classes into divisions (Table 2).

Form Genera

Generally, fossilized plants are more difficult to study than living plants because (1) most plant parts are not hard and thus are not readily preserved and (2) the plant usually breaks up into its component parts which are scattered before burial. Consequently, it is unusual to find plants fossilized with the stem, leaves, roots, and fruiting structures connected or even closely associated. Because of the difficulty of reconstructing the plant, the practice of giving separate names to each of the disassociated plant structures has developed. These names are called form genera and form species. An example is the naming of the Carboniferous treelike *Lepidodendron* (p. 51). The stem is referred to as *Lepidodendron,* the leaves as *Lepidophylloides,* the roots as *Stigmaria,* and the cones as *Lepidostrobus.* This complicated system of nomenclature has arisen from necessity. The close resemblance of some plant remains to those of other unrelated plants has been the principal reason for the use of form genera.

Geologic Distribution

With the exception of some algae and fungi, little is known about the plants of the Precambrian and Early Paleozoic eras (Table 3). These strata are principally marine; that is, they were deposited from sea water and, except for calcareous algae, contain very little preserved plant material. Early land plants have been reported from Upper Silurian time, and by Upper Devonian most major plant groups were present, including ferns, lycopods, sphenopsids, possible seed plants, algae, and fungi.

In the next two geologic periods, the Mississippian and Pennsylvanian, known collectively as the Carboniferous, the ferns, lycopods, calamites, seed-ferns, and cordaitean forms expanded in both numbers and development and the conifers appeared.

Floras of the Mesozoic era are composed mostly of the gymnospermous plant groups Cycadophyta, Ginkgophyta, Coniferophyta, and various ferns. The angiosperms or flowering plants had arisen at least by early Cretaceous, had become the dominant vegetational type by Upper Cretaceous, and have remained the dominant plant type to the present. Their development, one of the most rapid in the plant kingdom, was associated with the development of their pollinators, such as bees, wasps, and birds.

Table 3—Geologic Time Scale

Era	Periods	Epochs	Events of Plant Life	Million Years to Start of Period
Cenozoic	Quaternary	Recent		
		Pleistocene	Glaciation	1
	Tertiary	Pliocene	Rise of modern floras	11
		Miocene	Climates becoming drier over much of western U.S.	25
		Oligocene		40
		Eocene	Angiosperms continue to expand Climate in western U.S. subtropical to temperate	60
		Paleocene		70 ± 2
Mesozoic	Cretaceous		Expansion of flowering plants to worldwide dominance	135 ± 5
	Jurassic		Cycads and conifers dominant Ferns present	180 ± 5
	Triassic		Conifers increase, seed ferns decrease Ferns still relatively abundant. *Neocalamites* present Age of Petrified Forest N.P.	225 ± 5

Era	Period		Plant life	Age (million years)
	Permian		Treelike lycopods and *Calamites* decline and become extinct	270 ± 5
	Pennsylvanian	Carboniferous	Coal-forming swamps present in eastern U.S., not so much in the west Treelike lycopods, *Calamites,* seed ferns and *Cordaites,* and true ferns growing during this period	
	Mississippian			350 ± 10
Paleozoic	Devonian		Early land floras Lycopods, horsetail relatives, possible seed ferns, and progymnosperms present	400 ± 10
	Silurian		Land plants appear in Upper Silurian	440 ± 10
	Ordovician		Marine algae	500 ± 15
	Cambrian		Marine algae	600 ± 20
Precambrian			Primitive forms of algae and fungi	?

Types of Preservation and Methods of Study

Figure 26. Compression of angiosperm leaf

Fossil plants, although sometimes abundant locally, represent only a few of the total number of plants that have lived on the earth. After the plants died, most of them decayed, and only a few were preserved. Of those preserved, many were then subjected to such destructive natural forces as erosion. Thus, of the few initially buried plants, fewer survive as fossils, and of them only a comparatively tiny fraction are near enough to the earth's surface to be exposed or collected.

Plant remains have been preserved by many geologic processes. A common method is burial in sea, lake, or swamp sediments that later become rock. Ash fallen as a result of volcanic activity has literally entombed plants. In some areas of the world, plant fragments are preserved in calcareous nodules called coal balls and in amber (fossilized resin).

Fossil plants are most commonly preserved in three ways: (1) compressions or impressions, (2) casts or molds, and (3) petrifactions.

Compressions and Impressions

Compressions and impressions, particularly of leaves, occur commonly as fossils (Fig. 26). These formed when the remains of twigs and leaves fell into water, became soaked, settled to the bottom, and were covered by sediments. Overlying sediments accumulated, and their subsequent weight forced out the air and water, leaving only a thin layer of carbon. Frequently, these thin carbonaceous films of plant fragments

27

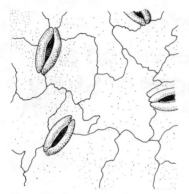

Figure 27. Cuticle, illustrating outlines of epidermal cells and stomata

contrast with the surrounding sediments, making beautiful specimens. When the rock is split, one-half of the specimen is nearly complete. That is the compression or positive side. Its counterpart is the impression or negative side.

The carbonaceous film may also contain the noncellular, waxy cuticle of the original leaf or stem. Cuticle, the protective outside covering of the epidermis, resists decay and may be removed from the rock for study under a microscope. The cuticle can be removed by applying three to five coats of Helen Neushaefer clear fingernail polish to the specimen. Each coat is applied after the previous application has dried. The coats should be allowed to dry for 24 hours and then peeled off. This peel can be studied under the microscope. The arrangement of the epidermal cells and stomata of such leaf cuticle is useful in the identification of some species (Fig. 27).

The external form of the plant fragment may have been modified by the weight of the overlying rocks. A round stem, for example, may become somewhat elliptical. Flat materials such as leaves are less affected.

Casts and Molds

A mold is formed in the rock when a cavity is left by decay of a stem, root, or seed. A cast results from the filling of this cavity, often by mud or sand. No organic material or cellular structure remains in a cast. Many stems and other solid parts are collected as casts and molds, such as *Lepidodendron* stems (Fig. 28) or calamitean pith casts (Fig. 57).

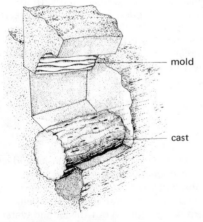

— mold

— cast

Figure 28. Cast of *Lepidodendron* stem

Figure 29. Petrified wood

Petrifactions

Petrifactions are mineralized plant remains (Fig. 29). They result when the wood or other structure is preserved by deposition of mineral material such as silica, pyrite, or calcium carbonate (calcite) along the cell walls. The foreign material either replaces or encloses the organic remains and, thus, often retains all the structural details of the original plant material. These types of fossils are very important botanically, because it is generally from petrified materials that the internal structure of the plant can be studied. One of the more famous petrified wood localities is the Petrified Forest National Park in Arizona. There, during the Triassic period or approximately 190 million years ago, trees that are relatives of some present-day conifers fell and were buried in mud. Water containing high concentrations of dissolved silica then percolated

29

Figure 30. Pseudofossils called dendrites

down into the logs and infiltrated spaces between and within the cells where the silica was deposited. Most of the plant material was replaced by the silica until only a small amount of the original plant organic matter remained.

The colors that make petrified woods so spectacular are caused by various oxides of iron and manganese. As an example, the red coloration may be due to small amounts of the iron oxide mineral called hematite (Fe_2O_3).

These petrified woods are studied by cutting and grinding portions of them into thin sections (p. 117). The cell structures of these translucent sections can then be studied under the microscope.

Pseudofossils

Pseudofossils are produced by the deposition of inorganic mineral matter, usually a black manganese oxide (pyrolusite), along cracks or other small openings between the bedding planes of the rock. These feathery-looking artifacts, called dendrites (Fig. 30), are often mistaken for the remains of ferns (Pl. 4, fig. 1).

Amateur collectors have mistaken rounded rocks for fossil apples, large seeds, and even the fingers of an ancient man!

Collecting and Processing Specimens

Where do I find fossil plants? This is largely dependent upon the specific locality and interest of the person asking the question. For beginning collectors, the place to start is near home; then, as they gain experience, they may range more widely. The best places for collecting are exposed sedimentary rocks in quarries or clay pits, road cuts, eroded surfaces, and shales associated with coals. The latter are more common in the Cretaceous of Utah and the Paleocene of Wyoming and Montana than elsewhere in the West.

The following table (4) indicates some localities. The locations are general, and more specific directions can be obtained locally.

Table 4—Some Collecting Sites

Quaternary
  Pleistocene   Not Common

Tertiary
  Pliocene
    Deschutes Formation   Along the Deschutes River southwest of The Dalles, Oregon

    Mount Eden Beds   Near Elsinore, California, northwest end of San Jacinto Mountains, California

  Miocene
    Latah Formation   Several road cuts around Spokane, Washington; south of Opportunity, Washington; east of Lewiston, Idaho

    Trout Creek flora   Eastern Oregon near Andrews, Oregon

31

Table 4 (cont'd)

Sucker Creek flora	Along Sucker Creek near Rockville, Oregon, and across the state line in Idaho
Mascall Formation	Road cuts east of Dayville, Oregon
San Pablo flora	North of Mt. Diablo near Clayton, California, near Altamont Pass east of Livermore, California;
Stinking Water flora	Road cuts near Burns, Oregon

Oligocene

Florissant flora	Near Florissant, Colorado. Much of this is included in Florissant Beds National Monument
Ruby flora	In Ruby Mountains, near Virginia City, Montana
Bridge Creek flora	Near Painted Hills State Park, Oregon
Weaverville flora	Along the Trinity River near Weaverville, California
Lower John Day flora	Behind school at Fossil, Oregon

Eocene

Green River Formation	Compressional flora near Bonanza, Utah; petrified woods at Eden Valley near Farson, Wyoming; the Blue Forest near Fontanelle, Wyoming
Salmon flora	Near city dump of Salmon, Idaho
Chalk Bluffs flora	Near Grass Valley, California
Copper Basin flora	Northeastern Nevada

Paleocene

Fort Union Formation	Several localities associated with coals in northeastern Wyoming and southeastern Montana. This formation extends northward into eastern Montana and western North Dakota, containing a number of localities.
Paskapoo Formation	Near confluence of Red and Blindman rivers, west of Red Deer, Alberta

Cretaceous

Upper

Mesa Verde Group	Northeastern Arizona and southwestern Wyoming around Evanston; Salina Canyon, Utah
Nanaimo Group	Near Comax and Nanaimo, British Columbia
Edmonton Formation	Forks of Michici Creek, near Drumheller, Alberta
Frenchman Formation	Southwest of Eastend, Saskatchewan

Middle

Dakota Formation	Northeast of Westwater, Utah

Lower

Cedar Mountain Formation	Near Moab and Castle Dale, Utah
Jackass Mountain Group	In the Frazier River valley near Lytton, British Columbia
Kootenay Formation	South of Blairmore and west of the old Blairmore Mine and the Mapleleaf Mine near Bellevue, Alberta

Spence Bridge Group	Near Spence Bridge on Thompson River, British Columbia
Jurassic	
Morrison Formation	Several localities near Hanksville and Moab, Utah; Greybull, Wyoming
Triassic	
Chinle Formation	Petrified Forest National Park (no collecting); several localities scattered throughout southern Utah, northeastern Arizona, and northwestern New Mexico
Permian	
Hermit Shale Formation	Grand Canyon National Park (no collecting)
Pennsylvanian	
Manning Canyon Shale	Lake Mountain, near Lehi, Utah

Figure 31. Collecting equipment

Figure 32. Leaf compression with label attached

Equipment necessary for collecting is inexpensive and readily obtainable. This equipment should include a geologist or bricklayer hammer (available at most hardware stores), a box or knapsack for carrying the specimens, chisels, a hand lens, both a road map and a topographic map for recording the exact location of the collecting sites, and a notebook for keeping a record of the specimen numbers and other information concerning the collections (Fig. 31).

In collecting, several principles need to be observed:

1. Collections should be as complete as possible. Sometimes several visits are necessary to obtain a representative collection of the plant types present. Small and fragmentary specimens, especially seeds and small leaves, should be retained along with the larger, conspicuous specimens. The more complete and better preserved the specimen is, the easier it is to identify and the more information it may yield.

2. To avoid mixing collections, the collector should number specimens consecutively and label them completely. This should be done at the first opportunity. The numbering can be done by placing a spot of white paint on the specimen, and, when it is dry, writing the number with a pen dipped in India ink. The number can also be written on the label (Fig. 32). The label should contain as much information about the specimen as possible. This is

33

easier than trying to remember after five years the name of the specimen and where it was collected. Because labels often become lost, the information on the label should also be kept in a separate book. Proper labeling enormously increases the potential scientific value of the specimen.

Information on the label and in the corresponding book should include: the number and the name of the specimen, when it was identified, who identified it, the precise geographic location of the collecting locality, the formation from which the specimen was obtained, and the name of the collector.

3. Specimens should be carefully packed and individually wrapped in paper to prevent damage caused by their rubbing together.

4. A good collection should be well organized. Organization can be by fossil type, that is, by petrifaction or compression, plant groups, flora, or any combination of these. The specimens should be kept in boxes and drawers for protection and ready accessibility. These may also be used as displays.

5. Specimens and collections can be presented to universities and museums, where they will be preserved and where paleobotanists can check their identifications, possibly bringing to light new scientific information. New specimens or localities brought to the attention of scientists have contributed significantly to the study of fossil plants.

Sequence of Plant Life of the Past

Floras of Paleozoic age in western North America have not been as extensively studied as those in the East. Fossil floras of this age are rare, primarily because widespread seas that covered much of western North America during Paleozoic time deposited thick sequences of marine sediments and, except for algae, plants were not generally preserved. Mesozoic floras are also relatively rare, whereas Cenozoic floras are fairly numerous. During the Tertiary period, volcanism provided favorable conditions for the preservation of plant remains in the volcanic ash and tuff that fell on the plants growing at that time.

Precambrian and Early Paleozoic

Fossil plants of the Precambrian are related to algae or fungi. These would include the stromatolites of Glacier National Park, Montana, and *Hadrophycus* from the Precambrian of the Medicine Bow Mountains, Wyoming.

Marine invertebrate fossils are common in the early Paleozoic, but plants other than algae are rare. Thus, the earliest fossil records in the West are evidences of algae such as various species of *Collenia, Autophycus,* and *Tetonophycus* described from the Cambrian Snowy Range Formation in the Teton Mountains, Wyoming.

On Beartooth Butte, near the northeast entrance of Yellowstone National Park, is a small Lower Devonian flora consisting of short axes of *Psilophyton wyomingensis, Hostinella* sp., and sporangia assigned to *Rebuchia* and *Broggeria.* These plants appear to have lived near lagoons on the edge of a Devonian sea (Fig. 33).

35

Figure 33. Generalized Devonian landscape reconstruction

Chart 1—Distribution of Genera in Carboniferous Floras of Western North America

	Honaker Trail (?) flora, Utah	Mosquito Range flora, Colorado	Manning Canyon Shale flora, Utah	Salida Canyon flora, Colorado	Spotted Ridge flora, Oregon
Asterophyllites *	●	X	X	X	X
Annularia *	X	●	●	●	●
Archaeocalamites *	●	●	X	●	●
Artisia **	X	●	X	●	●
Aulacotheca ++	●	●	X	●	●
Calamites *	X	X	X	X	●
Cardiocarpus +++	●	X	X	●	●
Cordaianthus **	●	X	X	●	?
Cordaites **	X	X	X	X	●
Cornucarpus +++	●	●	X	●	●
Corynepteris *** (Alloiopteris)	●	●	X	●	●
Crossopteris ***	●	?	X	●	●
Gnetopsis +++	●	●	X	●	●
Lepidocarpon +	●	●	X	●	●
Lepidodendron +	●	X	X	●	●

36

Figure 34. Reconstruction of central Utah during Carboniferous period

Lepidophloios +	X	•	X	•	•
Lepidophylloides +	•	•	X	•	•
Lepidostrobo-phyllum+	•	•	X	•	•
Lepidostrobus +	•	X	X	•	•
Mesocalamites *	•	•	X	•	X
Neuropteris ***	?	X	X	X	•
Odontopteris ***	X	•	X	X	•
Palaeostachya *	•	•	X	•	•
Pecopteris ***	X	•	•	X	X
Rhodea ***	•	•	X	•	•
Rigbyocarpus +++	•	•	X	•	•
Sigillaria +	•	•	X	•	•
Sphenopteridjum ***	•	•	X	•	•
Sphenopteris ***	•	X	X	•	•
Stigmaria +	•	X	X	•	X
Trigonocarpus +++	•	•	X	•	•
Walchia *+	•	•	•	X	•

Calamitales *, Cordaitales **, fernlike foliage ***,
Lepidodendrales +, microsporangiate structures ++,
seeds +++, Coniferales *+

37

Late Paleozoic

Fossil plants of late Paleozoic age have been studied extensively throughout the world. The similarity among these floras is striking. They were more uniform in composition than any other forms at any time during the history of the earth. This uniformity is attributed to the absence of well-defined climatic zones. In conditions of perpetual summer, forests of fast-growing, rather soft-tissued "trees" spread over the moist lowlands of the Carboniferous landscape, dominating the world. Fossil floras illustrating the swampy, lowland conditions of this period are the Manning Canyon Shale flora of central Utah (Fig. 34), the Pennsylvanian Spotted Ridge flora of central Oregon, and the Mississippian Uintah flora of northeast Utah. The flora from the Lower Pennsylvanian Manning Canyon Shale is highly diversified and contains such forms as *Lepidodendron, Sigillaria* and their related genera; jointed, ribbed stems with whorled leaves assigned to *Calamites* and *Asterophyllites* respectively; fernlike foliage such as *Sphenopteris, Rhodea, Neuropteris, Crossopteris,* and *Corynepteris;* many isolated seeds; and the remains of cordaitean plants, particularly their straplike leaves *(Cordaites)* and rather loosely compacted cones *(Cordaianthus).*

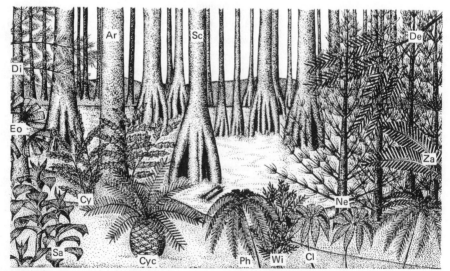

Figure 35. Generalized reconstruction of Triassic landscape. Ar *Araucarioxylon,* Cl *Clathropteris,* Cy *Cynepteris,* Cyc *Cycadeoidea,* De *Dechellyia,* Di *Dinophyton,* Eo *Eoginkgoites,* Ne *Neocalamites,* Ph *Phlebopteris,* Sa *Sanmiguelia,* Sc *Schilderia,* Wi *Wingatea,* Za *Zamites.*

Near Moab, Utah, a flora of possible Upper Pennsylvanian age has been reported. This flora contains lycopods, *Annularia, Odontopteris,* and pecopterid leaves.

Extensive mountain building during the Permian brought about drastic changes in the environments, and plants and animals changed accordingly. Lycopods, *Calamites,* and several other Pennsylvanian "holdovers" common at the first of the period did not adapt and therefore became very rare or extinct. True conifers became dominant, and early cycadlike forms appeared in the Permian. Species of the Hermit flora in the Permian of the Grand Canyon region in northern Arizona appear less abundant than in floras of corresponding age elsewhere in the world. The flora suggests a delta plagued by showers, burning sun, occasional torrents, periods of drought, and drying pools (White, 1929).

Early Mesozoic

Conifers were one of the most important plant groups in early Mesozoic floras. During the Triassic a shallow sea covered much of the Southwest. Later in the period a large flood plain developed in northern Arizona and southern Utah, where many conifer-type logs were deposited along with foliage of ferns and cycadeoids. *Araucarioxylon, Woodworthia,* and other plants were preserved in what is now Petrified Forest National Park in east-central Arizona (Fig. 35).

Jurassic seas occurred widely in the western United States and Canada. Late Jurassic streams from uplifting mountains spread a sheet of mud and other sediments across the lowlands from northern Montana to southern New Mexico, resulting in what eventually became the Morrison Formation. Dinosaurs were trapped in the muds of this formation and their bones petrified. Plant remains are also abundant in this formation and include conifer wood, cones, short shoots, and leaves; seeds; cycadeoid stems and leaves; *Osmundacaulis;* and other forms. Jurassic forests appear to have been predominately conifers associated with cycadeoids and an understory of ferns.

Late Mesozoic

Modern western floras had their beginning in the Cretaceous. Gymnosperms, particularly conifers,

Chart 2—Distribution of Genera in Some Mesozoic Floras of Western North America

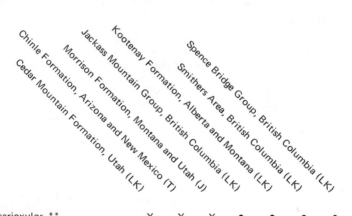

Genus	Cedar Mountain Formation, Utah (LK)	Chinle Formation, Arizona and New Mexico (T)	Morrison Formation, Montana and Utah (J)	Jackass Mountain Group, British Columbia (LK)	Kootenay Formation, Alberta and Montana (LK)	Smithers Area, British Columbia (LK)	Spence Bridge Group, British Columbia (LK)
Araucarioxylon **	X	X	X	•	•	•	•
Behuninia †	•	•	X	•	•	•	•
Cephalotaxis **	•	•	•	•	?	•	•
Cladophlebis ++	•	X	•	X	X	X	X
Clathropteris ++	•	X	•	•	•	•	•
Coniopteris ++	•	•	•	X	X	X	•
Ctenis ***	•	•	•	•	X	•	•
Cycadeoidea +	X	X	X	•	•	•	•
Cynepteris ++	•	X	•	•	•	•	•
Equisetum (horsetail) ††	•	•	•	•	X	•	•
Gleichenia ++	•	•	•	X	•	•	X
Jensensispermum †	•	•	X	•	•	•	•
Metasequoia (dawn redwood) **	•	•	•	?	•	?	•
Monanthesia +	X	•	•	•	•	•	•
Neocalamites *	•	X	•	•	•	•	•
Nilssonia ***	•	•	•	X	X	•	X
Osmundacaulis ++	•	•	X	•	•	•	•
Paraphyllanthoxylon **	X	•	•	•	•	•	•
Phlebopteris ++	•	X	•	•	•	•	•
Pityocladus**	•	•	X	•	•	•	•
Pityophyllum**	•	•	X	•	•	X	X
Podozamites**	•	•	X	•	X	•	X
Protopiceoxylon**	•	•	X	•	•	•	•
Pseudoctenis***	•	•	•	•	X	•	•
Pseudocycas +	•	•	•	•	•	X	X
Pterophyllum +	•	•	•	X	•	•	•

Ptilophyllum +	•	•	•	X	X	X	•
Sequoia (redwood) **	•	•	•	•	•	•	?
Sphenopteris +++	•	•	•	X	X	X	X
Wingatea ++	•	X	•	•	•	•	•
Xenoxylon **	•	•	X	•	•	•	•
Zamites +	•	X	•	•	X	•	•

Calamitales*, Coniferales **, Cycadales ***, Cycadeoidales +, Ferns ++, Fernlike foliage +++, seeds †, Equisetales ††

(J = Jurassic, LK = Lower Cretaceous, T = Triassic)

Figure 36. Generalized reconstruction of central Utah during Upper Cretaceous Blackhawk Time

were dominant in Lower Cretaceous floras, but their numbers and importance diminished because of competition from the rapidly spreading flowering plants.

As in the Jurassic, cycadeoids were growing on the upland slopes during the Cretaceous. They occur in the Lower Cretaceous flora of the Cedar Mountain Formation in east-central Utah and the Burro Canyon Formation of southwestern Colorado and are associated with *Tempskya,* petrified conifer logs and fossil angiosperm wood. The angiosperms are *Paraphyllanthoxylon* and *Icasinoxylon.*

The Middle Cretaceous Dakota flora near Westwater, Utah, is an admixture of ferns and angiospermous species. The flora is dominated by the ferns *Astralopteris, Matonidium, Gleichenia,*

41

Asplenium (spleenwort), and *Coniopteris.* The angiosperms are *Ilex* (holly), *Ficus, Eucalyptus, Platanus* (sycamore), and *Salix* (willow).

Late Cretaceous floras indicate a warm-temperate to subtropical climate. Genera reported from the Medicine Bow flora of south-central Wyoming include *Ficus, Dryophyllum, Cinnamomum, Laurus, Magnolia, Sabal* (palm), *Metasequoia* (dawn redwood), and several other forms. Genera described from this flora suggest a heavy rainfall distributed throughout the year without any frost.

Flora of the Upper Cretaceous Blackhawk Formation at Salina Canyon in central Utah appears to have grown under warm, humid lowland conditions (Fig. 36). The flora consists of two palm genera: *Sabalites* and *Phoenicites (=Genomites);* species of the ferns: *Allantodiopsis, Saccoloma,* and *Osmunda;* and several conifers: *Sequoia* (redwood), *Protophyllocladus,* and *Araucaria* (Norfolk Island pine). Among other angiosperms in this flora are *Dryophyllum, Juglans* (walnut), *Cercidiphyllum, Magnolia, Menispermites, Ficus,* and *Salix.*

Cenozoic

Tertiary Period

The climate of the West during the Tertiary period and extending into the Quaternary was dominated by three trends: (1) a general cooling trend, often interrupted by warmer periods, (2) a progressive drying from the early Miocene into the Pleistocene because of increasing uplift of the Sierra Nevada and Cascade Mountains, and (3) during the Pleistocene, alternating glacial and interglacial stages.

The vegetation pattern of the Tertiary was marked by extensive "migrations" of plant groups in response to these climatic changes. These migrations were basically a withdrawal of tropical and semitropical vegetation southward as the climate of the continents grew cooler and drier in response to the rising mountain ranges.

Paleocene Epoch. Strata of Paleocene age in Montana, eastern Wyoming, and the western Dakotas, collectively called the Fort Union Formation, contain large amounts of mineable coal. Associated with these coals is a diversified flora that includes algae, fungi, bryophytes, ferns and their allies, conifers, cycads and ginkgos. The

42

Figure 37. Eocene landscape reconstruction, Chalk Bluff, California

highly varied angiosperms include palms, willows, maples, ash, *Arctocarpus* (breadfruit), and *Zelkova* (keaki tree) among many others.

The Fort Union flora was a lowland flora that probably grew in a subtropical to warm temperate climate and extended inland from the Paleocene Sea to the mountains. A general overview of the Fort Union flora indicates that many of the species likely occurred on drier hillsides some distance from the swampy lowlands and that leaves, fruits, seeds and stems were borne downstream to lower points of deposition.

Eocene Epoch. Eocene floras found in the Arctic contain several forms similar to those currently growing in Virginia and the Carolinas. This suggests a rather mild climate for the Arctic during this epoch.

During the Eocene, several large lakes existed and most of the western United States was near sea level. The western half of the Great Basin and the Sierra Nevada beyond was a low upland. The lakes, known as the Green River System, covered approximately 50,000 square miles of what is now northeastern Utah, southwestern Wyoming, and northwestern Colorado. Accumulated in these lakes are deposits, called the Green River Formation, that contain a large number of well-preserved fossil plants, insects, fish, and mammals. Described plant genera from the Green

River Formation include such woodland types as *Celtis* (hackberry), *Pinus* (nut pine), *Platanus* (sycamore), *Rhus* (sumac), *Vauquelinia,* and such subtropical forms as *Cardiospermum* (balloon vines), *Mimosites,* and *Thouinia.*

The Middle Eocene Salmon flora is found in lake beds near Salmon, Idaho. This flora has a rich representation of ferns (*Osmunda, Woodwardia,* and *Adiantum),* numerous rosaceous shrubs, and an abundance of conifers, all of which point to a cool-temperate climate with ample precipitation.

The Upper Eocene Copper Basin flora preserved near Jarbidge, Nevada, contains an unusual number of montane conifers, many rosaceous shrubs, and flowering plants with comparatively small leaves. The rarity of broad-leafed deciduous hardwoods and the near absence of mild-temperate species in general suggests that this flora was a conifer-hardwood forest close to a mountain conifer forest. Along the borders of streams and lakes were dense groves of *Alnus* (alder), *Acer* (maple), and *Lithocarpus.* Shrubby plants such as *Amelanchier* (serviceberry), *Crataegus* (hawthorn), *Mahonia* (Oregon grape), *Prunus, Ribes,* and *Salix* (willow) are common. The conifer-hardwood forest that occurred in the bordering valleys and lower slopes consisted of *Cephalotaxus, Pseudotsuga* (douglas fir), *Chamaecyparis* (white cedar) and *Sequoia* (redwood). These were associated with hardwoods such as *Acer, Arbutus, Prunus,* and *Alnus.* The common understory plants were *Sassafras, Amelanchier, Crataegus, Prunus,* and *Ribes.*

It is postulated that the Eocene climate of the Pacific Northwest lowlands was subtropical and, thus, similar to the present climate of Costa Rica in Central America. The Clarno flora of north-central Oregon contains species of palms, cycads, evergreen oaks, figs, laurel, and similar broad-leafed evergreen plants, as well as *Osmunda* rhizomes (Fig. 229).

Oligocene Epoch. This epoch began with subtropical to warm-temperate conditions but later changed to a more temperate climate. This was the result of the cooling and drying trend that continued progressively—with brief recessions to warmer climates—through the remainder of the Tertiary period.

One of the most widely known Oligocene floras in the western United States is in the Florissant beds near Colorado Springs, Colorado

44

(Pl. 21). It occurs in thinly bedded strata consisting mainly of pumice and volcanic dust that were deposited in an ancient lake. Plants of this flora appear to have been overwhelmed by the volcanic ash. Remains of deciduous trees, palms, stumps of *Sequoia,* and fauna such as insects were buried beneath the ash. One hundred fourteen species of plants have been classified from this flora. The most abundant species, based on the number of specimens collected, is *Fagopsis longifolia* (p. 111), which comprises 30 percent of the total. Some genera no longer native to North America are *Lomatia, Zelkova* (keaki tree), *Ailanthus* (tree of heaven), and *Koelreuteria* (goldenrain tree). The flora appears to have grown at an elevation between 1,000 and 3,000 feet. The climate was perhaps warm temperate to subtropical, with enough rainfall to support a large forest growth along the stream banks. Pines and oaks dominated the open fields on the higher ground.

The Ruby flora occurs in strata of late Oligocene age on the eastern slopes of the Ruby Range in southern Montana. This flora occurs in extensive lake deposits and appears to have been a rather dry coniferous-deciduous association without any subtropical plants. The leaves are in general smaller than average, which suggests reduced precipitation. The Ruby flora contains species of such genera as *Abies* (fir), *Picea* (spruce), *Pinus* (pine), *Metasequoia* (dawn redwood), *Acer* (maple), *Betula* (birch), *Celtis* (hackberry), *Fagus* (beech), *Fraxinus* (ash), *Mahonia* (Oregon grape), *Populus* (poplar), *Ribes, Rosa* (rose), *Ulmus* (elm), and *Zelkova* (keaki tree).

Miocene Epoch. During the Miocene and Pliocene epochs, the climate became more diversified in the western United States. Species of floras of this geologic age are close enough to modern forms to suggest that they lived in an environment having a similar altitude and rainfall. They indicate that the Sierra Nevada mountains cast an ineffective rain shadow over the West during the Miocene period. These mountains are postulated to have been less than 3,000 feet above sea level at that time, and the Great Basin only 1,000 feet lower.

Some of the better known floras of Miocene age are: in Oregon—Mascall (Fig. 38), Stinking Water, Sucker Creek, and Trout Creek; in Idaho—Thorn Creek; in Washington—Latah; and in Nevada—Fingerrock.

Figure 38. Reconstruction of bald cypress (*Taxodium*) swamp in central Oregon during the Miocene (Mascall flora)

Components of these floras indicate that by early Miocene the subtropical forest had moved southward and a temperate forest dominated by *Sequoia* and associated plants occupied much of Oregon and the Great Basin.

Herbaceous angiosperms increased rather distinctly during the Miocene. It is interesting to note that *Artemesia tridentata,* commonly known as sagebrush, first occurred in the upper Miocene and presently forms the climax cover over much of the higher cool deserts of the West.

By the end of the Miocene, moisture had decreased in the northern portion of the Great Basin. Eastern Oregon became drier, and a comparatively dry woodland flora occupied western Nevada. This resulted from further rising of the Cascades and the northern Sierra Nevada, which formed a more prominent rain shadow.

Pliocene Epoch. The Sierra Nevada and Cascade mountains underwent their greatest uplift during the Pliocene, exerting a strong climatic influence over the area to the east. This uplift extended into the Pleistocene until the Sierra Nevada block rose from 5,000 to 6,000 feet in the north and from 7,500 to 9,000 feet in the south, thus casting a rain shadow over much of the West.

Floras from the Pliocene contain species that indicate a close relationship to trees presently growing in western North America. The Pliocene

46

Figure 39. Reconstruction of Ice Age (Pleistocene Epoch)

is thought to be the record of an immediate past and not of remote yesterdays. Although plant life was probably as abundant and varied as during the Miocene, the megaplant fossil record of the Pliocene and Pleistocene is scanty when compared with those of other epochs of the Tertiary Period. The scarcity of these fossil plants appears to have been caused by the dryness of the habitats, a condition unfavorable for plant preservation.

Temperatures in late Pliocene were very much like those of today, although the trend was toward cooler temperatures and eventually culminated in the ice ages of the Pleistocene. Many earlier plant forms in western North America were eliminated by climatic changes. Late Pliocene floras were adapted to less than 15 inches of annual rainfall; and many floras had changed into savanna and grassland floral types, with maples, poplars, *Prunus,* and willows growing along the streams.

Quaternary Period

Pleistocene Epoch. The Pleistocene or the Ice Age, which consisted of four major glacial advances, is thought to have covered some 1 million years. Well-preserved glacial features such as moraines and cirques are readily observed demonstrating the extent of glaciation in the West (Fig. 39). Several basins, whether they

47

presently contain water or not, repeatedly held lakes. Lake Bonneville, which was situated in western Utah, was nearly the size of Lake Michigan; and Lake Lahontan in northwestern Nevada was similar in size to Lake Erie or Lake Ontario. Lake Bonneville attained an elevation about 1,000 feet above Great Salt Lake, and Lake Lahontan reached a level some 515 feet above present Pyramid Lake.

A temperature difference of only 4.5-5.4° F separated the overall temperature of the Upper Wisconsin ice age of the Pleistocene and the overall temperature of the present Great Basin. This was probably the result of a warming trend of the Pacific Ocean.

The last glacial period was about 12,000 years ago, at which time Lake Bonneville overflowed into the Snake River. As the climate became drier, glaciers waned, streams dwindled, and lakes evaporated faster than streams could fill them.

Vegetation fluctuated in response to alternating moist and dry climates during the Pleistocene epoch. At present, the climate is in a dry phase and desert floras cover much of the West.

Algae

Figure 40. Longitudinal section of *Cryptozoon* stromatolite

Figure 41. Longitudinal section of *Collenia* algae

Algae are the least complicated of all the green plants. The body, known as a thallus, generally lacks characteristics that allow for fossilization. Only algae that cause the precipitation of carbonates from surrounding water and those that actually secrete carbonate or silica in their cell walls are likely to be fossilized after burial. Consequently, few specimens in the fossil record can be directly related to algae.

Cyanophyta

As far as can be interpreted from the fossil record, algae and fungi were the principal forms of Precambrian plant life. These Precambrian algae are represented by large, laminated calcareous deposits closely resembling algal reefs and known by the informal term *stromatolites.* The cellular structure of these crusts is rarely preserved. They are common in the Precambrian Belt Series of Glacier National Park and elsewhere.

Stromatolites are classified on the basis of their growth habit and colony form. The two most common types are *Cryptozoon* (Fig. 40) and *Collenia* (Fig. 41). They are very much alike. *Cryptozoons* are depressed hemispheroidal or turbinate (inverted cone-shaped) colonies of variable outline (Pl. 22, fig. 1). *Collenia* consists of erect, columnar or pyriform (pear-shaped) colonies that are cylindroidal or hemispheroidal in outline (Pl. 22, fig. 5). The latter are often composed of alternating resistant and weak layers. *Collenia* has digitate (fingerlike) or

49

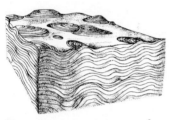

Figure 42. *Codonophycus*

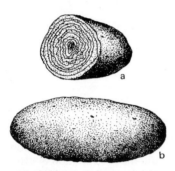

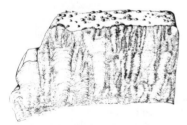

Figure 43. Algal balls. a. Algal ball with cutaway section, showing concentric rings, b. Entire ball

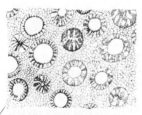

Figure 44. Longitudinal section of *Chlorellopsis*

turbinate extensions from the upper surface. These are not developed in *Cryptozoon*.

Codonophycus

This genus has been reported from rocks of Mississippian age in the Big Horn Mountains of Wyoming. It contains bell-shaped, conical or columnar, laminated algal colonies connected to each other by convex or undulant areas (Fig. 42; Pl. 39, fig. 5).

Algal Balls

Balls up to softball size have been collected from the Flagstaff Limestone Formation (Eocene) of Utah (Pl. 22, fig. 4). These are made up of concentric wavy or crinkly rings of calcite that many geologists consider to be algal secretions (Fig. 43), although no definite algae have been observed.

Chlorophyta (Chlorophycophyta)

Chlorellopsis

Chlorellopsis forms most of the algal "reefs" of the Eocene Green River Formation. These algae also encrust many of the petrified stems in this formation from Eden Valley, Wyoming.

The reefs take a variety of forms, the huge, puffball-shaped heads being most typical (Pl. 22, fig. 2). These algae also occur as isolated elliptical heads (Fig. 44).

In microscopic view, *Chlorellopsis* illustrates a spherical shell of very fine interlocking grains of calcite. The cellular structure is preserved where the calcite has entered through a rupture in the wall of the cell (Fig. 45).

Figure 45. Enlarged spherical cells of *Chlorellopsis*

✓Lycopodophyta

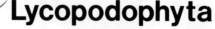

Figure 46. Reconstruction of *Lepidodendron*

(*Lepidophyllum*)

Lepidodendrales

The Lepidodendrales comprise a very important part of Carboniferous floras. They were mostly treelike, and the structure of their rootlets and of the aerating tissues in their stems indicates that for the most part they grew in swamps of fresh or brackish water.

The fossil-stem genera *Lepidodendron* and *Sigillaria* are ancient relatives of the small, living forms *Lycopodium, Selaginella,* and *Isoetes* and resemble these living forms in many ways, particularly in fruiting structures.

The stems of the fossil genus *Lepidodendron* (Fig. 46) are covered by characteristic scalelike scars (leaf cushions) (Fig. 47) that persist after long, linear, needle-shaped leaves *(Lepidophylloides)* (Pl. 2, fig. 1) have fallen off. Hence, the name from the Greek: *lepis, "scale"; dendron, "tree."*

The leaf-cushion on *Lepidodendron* is somewhat diamond shaped and has a small, transverse leaf scar in the upper part, where the leaf was actually attached (Pl. 24, fig. 6). In the center of the leaf scar, three dots may be observed. The center one is the small vascular scar where the conducting strand (midvein) of the leaf connected with the conductive tissue in the stem. The two scars on each side, called parichnos, represent tissue for the exchange of oxygen and carbon dioxide. Two more scars that may be present beneath the leaf scar represent the remains of the aerenchyma tissue, which also functions in gaseous exchange.

51

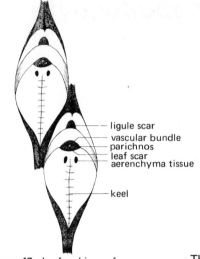

- ligule scar
- vascular bundle
- parichnos
- leaf scar
- aerenchyma tissue

- keel

Figure 47. Leaf cushions of
Lepidodendron

Figure 48. Reconstruction of
Sigillaria

The genera *Lepidodendron* and *Sigillaria* (Fig. 48) are distinguished from one another by the arrangements of the leaf-cushions on their respective stems. The cushions are arranged spirally in *Lepidodendron* (Fig. 47) and vertically in *Sigillaria* (fig. 49; Pl. 24, figs. 4 and 5).

Another genus closely related to *Lepidodendron* is *Lepidophloios* (Fig. 50). It differs in having spirally arranged leaf-cushions that are elongated horizontally and have the leaf scar near the base of the cushion on older stems (Pl. 3, fig. 3), whereas the cushions in *Lepidodendron* are vertically elongated.

The cones of *Lepidodendron,* placed in the genus *Lepidostrobus* (Fig. 51), grew attached to the tips of the smaller branches when the plant was living but became disassociated from the tree at death. The cones were cylindrical, although they are generally preserved as compressions (Pl. 24, fig. 2). They consist of a central axis that bears crowded spirals of bracts (sporophylls) with sporangia on their upper surface. Most cones contain two types of spores, the larger megaspores near the base and the small microspores near the cone apex.

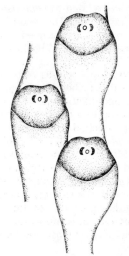

Figure 49. Leaf cushions of
Sigillaria

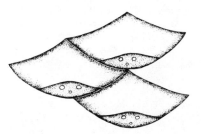

Figure 50. Leaf cushions of
Lepidophloios

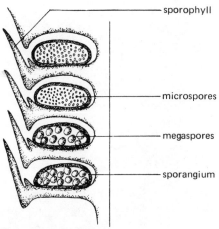

Figure 51. *Lepidostrobus*

53

Some cones, closely related to *Lepidostrobus,* developed seedlike structures. These cones and their detached sporophylls are assigned to the form genus *Lepidocarpon* (Fig. 52). The cones are thought to have disintegrated at maturity, releasing the individual sporophylls with their seedlike sporangia still attached (Pl. 3, fig. 4). The occasional sporophylls that were flattened during preservation and that can be collected with the sporangium still attached are assigned to the genus *Lepidostrobophyllum* (Pl. 23, fig. 6).

Stigmaria (Fig. 53) is the name given to the rhizophore or rootlike system belonging to *Lepidodendron, Sigillaria,* and *Lepidophloios.* Stigmarian remains with the rootlets attached are often found where they grew (Pl. 24, fig. 1). They generally occur preserved as compressions or casts showing the spirally arranged rootlet scars.

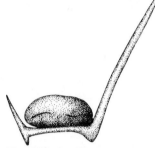

Figure 52. *Lepidocarpon*

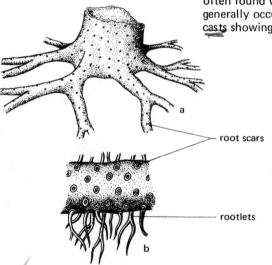

root scars

rootlets

Figure 53. *Stigmaria.* a. Reconstruction of rhizophore, b. Closeup, showing rootlet scars and attached rootlets

⁄Sphenophyta

Figure 54. Reconstruction of Calamites

Figure 55. Asterophyllites

Calamitales

The Carboniferous relatives of the living *Equisetum* or joint grass were numerous and varied. While most were treelike, a few were herbaceous. The treelike forms called *Calamites* lived in swamps (Fig. 54), and had creeping, underground stems (rhizomes) that probably produced a rather dense, junglelike growth.

Because of the difficulty of piecing together the fossilized remains of leaves, stems, and cones, they are given separate generic (form-genera) and specific names.

Asterophyllites Fig. 55; Pl. 3, fig. 1) or *Annularia* (Fig. 56, Pl. 3, fig. 2), were borne successively in whorl after whorl at the nodes or joints of the stems or branches of *Calamites*. Asterophyllitean leaves are of equal length and are cupped around the axis, whereas *Annularia* are flattened and resemble a small flower. Calamitean forms have been reported from the Manning Canyon Shale of Utah and the Spotted Ridge Formation of Oregon.

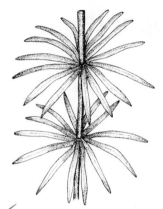

Figure 56. *Annularia*

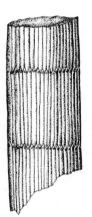

Figure 57. Pith cast of *Calamites*

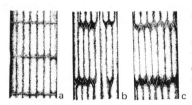

Figure 58. Calamitean genera represented by compressions of pith casts. a. *Archaeocalamites,* b. *Mesocalamites,* c. *Calamites*

Generally, the external portion of the *Calamites* stem is not preserved. Those structures that are fossilized, referred to as pith casts (Fig. 57), consist of longitudinally ribbed and grooved compressions of the pith of the living plant. The pith disintegrated while the plant was living and, at its death, filled with sediments. The remainder of the plant was destroyed, leaving the impression of the inner surface of the woody cylinder. The internodal grooves indicate the impressions of wood or xylem wedges projecting into the pith cavity. These wedges were composed of the primary water-conducting tissues of the plant. The raised ribs between the furrows mark the position of the rays.

Two genera and one subgenus, defined according to whether the ribs alternate or pass directly through the nodes, have been reported from the Manning Canyon Shale and the Spotted Ridge floras. Specimens in which all the ribs alternate are true *Calamites* (Fig. 58c; Pl. 23, fig. 2); those with some ribs passing directly through the nodes and some ribs alternating are assigned to the subgenus *Mesocalamites* (Fig. 58b; Pl. 23, fig. 1); and those with all their ribs passing directly through the nodes are termed *Archaeocalamites* (Fig. 58a; Pl. 23, fig. 4).

The nodes often bear circular scars (Fig. 59) caused by the breaking off of branches.

Calamitean fruiting structures consist of small, narrow cones at the end of each slender branch. Although many cone genera have been described for *Calamites,* only one, *Calamostachys,* has been reported from the western United States (Manning Canyon Shale). In this genus, the narrow cones are composed of alternating whorls of fertile structures (sporangiophores) and sterile bracts (Fig. 60). Sporangiophores bear the spore-producing sporangia. In compressions the bracts are readily observed, whereas the fertile portion is generally obscure.

Palaeostachya, another cone genus assigned to *Calamites,* has been collected from the Manning Canyon Shale (Pl. 23, fig. 3). In *Palaeostachya* the sporangia and sporangiophores arise at an angle from the axil of the bract (Fig. 61), whereas in *Calamostachys* (Fig. 60), they are perpendicular to the cone axis above the bract.

Neocalamites is a Triassic genus that resembles *Calamites.* The markings on the surface of the pith cast of *Neocalamites* are broad, curved, shallow depressions that pass directly through the nodes. The leaves, which are rarely preserved,

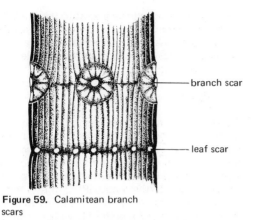

branch scar

leaf scar

Figure 59. Calamitean branch scars

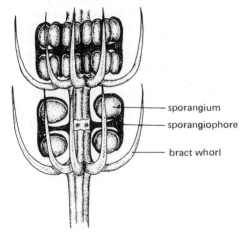

sporangium

sporangiophore

bract whorl

Figure 60. Enlarged portion of *Calamostachys* cone

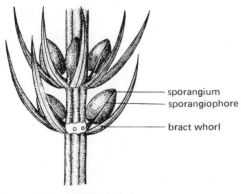

sporangium
sporangiophore

bract whorl

Figure 61. Enlarged portion of *Palaeostachya*

Figure 62. *Neocalamites* leaves

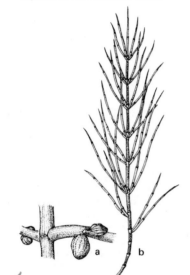

Figure 63. *Equisetum* (horsetail).
a. Bulbils on rhizome, b. Stem
with branches in whorls

are broadly attached and widest in the central
portion, tapering gradually to a point (Fig. 62).
This genus has been collected from the
Triassic Dolores Formation and the Chinle Formation, of Colorado and Utah respectively.

Equisetales

Equisetales is an order represented by only one
living genus, *Equisetum*, and its fossil counterpart *Equisetites*. *Equisetum*, (Fig. 63b), commonly called horsetail or joint grass, has jointed
stems, whorled leaves, and branches with a terminal cone on fertile shoots (Pl. 2, figs. 5 and
6). Upright and underground stems (rhizomes)
similar to those of *Equisetum* have been found in
the fossilized state and are sometimes called
Equisetites. The upright portion of fossil *Equisetum* looks like modern *Equisetum.* The
rhizomes of *Equisetum* are horizontal and
smooth to slightly ribbed, bearing rounded,
wrinkled tubers or bulbils (Fig. 63a; Pl. 2, fig.
4). *Equisetites* or *Equisetum* has been reported
from the Carboniferous to the Recent from
various localities around the world.

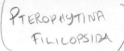

PTEROPHYTINA
FILICOPSIDA

Filicophyta (Ferns)

The true ferns span a wide geologic range, extending from the Recent back to the Middle Devonian. The number of fern species since the Carboniferous Period has remained rather constant. Although the ferns have never formed a dominant vegetational type during any period in geologic history, they were a conspicuous component of floras of the Upper Paleozoic and Mesozoic eras.

The leaf portion of a fern, called the frond, consists of two parts: the stalk, or petiole, and the blade (Fig. 64). The blade may be simple but most commonly is compound. Compound leaves, made up of many distinct smaller segments or leaflets called pinnules, may be either pinnately or palmately compound. In a pinnately compound leaf (Fig. 65), the petiole extends as the rachis (main axis) of the leaf from which the primary lateral segments of the blade arise. Further dissection of the leaf results in a bipinnate, tripinnate, or quadripinnate leaf. The lateral branches of a pinnately compound leaf are called pinnae—either primary, secondary, or tertiary, depending on the amount of dissection. In a palmately compound frond (Fig. 78), the segments or pinnules are all connected to the petiole at a common point, in the same way that the fingers are attached to the hand. One of the distinguishing characteristics of the fern is the coiled, immature frond, known as the fiddlehead.

The stems of most ferns, which grow horizontally underground and are called rhizomes, produce a new set of leaves each growing season. Some tree ferns have an aerial stem that may

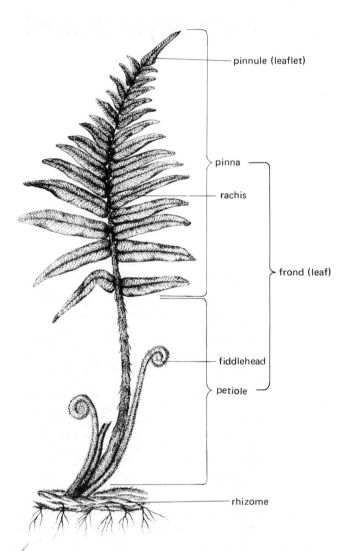

pinnule (leaflet)

pinna

rachis

frond (leaf)

fiddlehead

petiole

rhizome

Figure 64. Morphology of pinnately compound fern frond

reach 30 to 45 feet in height and up to 2 feet in diameter.

Ferns reproduce by means of spores. Usually the sporangia are grouped together on the lower surface of leaves into structures called sori or fruit dots (see *Matonidium*). Often the sori have a protective covering known as an indusium and thus are said to be indusiate (Fig. 66). Those that lack the indusium are referred to as nonindusiate.

Fern fronds are found preserved in many formations as compressions. Following are descriptions of some common fern genera identified from fossil floras of western North America.

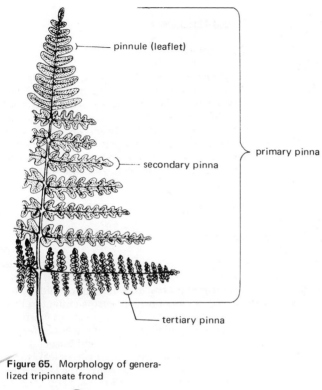

pinnule (leaflet)

primary pinna

secondary pinna

tertiary pinna

Figure 65. Morphology of generalized tripinnate frond

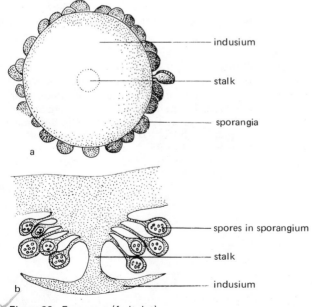

indusium

stalk

sporangia

a

spores in sporangium

stalk

indusium

b

Figure 66. Fern sorus (fruit dot).
a. Bottom view, b. Side view

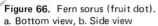

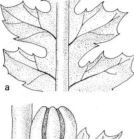

a

b

Figure 67. *Corynepteris.* a. Sterile pinnules, b. Pinnules with attached sori

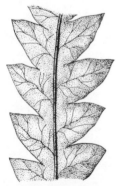

Figure 68. *Osmunda*

Zygopteridaceae

Corynepteris

Corynepteris (Fig. 67) is a genus for fertile pinnules and is assigned to the true ferns. The very small, sterile pinnules are broadly attached (sessile) and have roughly a square outline with three or four pointed teeth (Pl. 26, fig. 3). Five or six sporangia are grouped to form a spherical sorus. Sterile leaves similar to *Corynepteris* and assigned to the form genus *Alloiopteris* commonly occur in the Manning Canyon Shale flora.

Osmundaceae *primitive*

Osmunda

Osmunda leaves occur in Cretaceous and Tertiary floras. In the western United States this genus was restricted in distribution during mid-Tertiary, and subsequently it became extinct in this region. Currently it is living in eastern North America. (Members of this family also grow in South America, Africa, Asia, and Europe.)

Osmunda is a delicate fern whose leaf is twice pinnate (Pl. 6, fig. 2). The veins in the lanceolate pinnules are generally well marked and consist of a rather thick midvein and twice forked, fine lateral veins that arise obliquely (Fig. 68).

Cladophlebis

Cladophlebis (Pl. 7, fig. 1) is a form genus for sterile, linear to sickle-shaped, entire-margined pinnules, the petioles of which have never been observed. The pinnules have sparse venation and a distinct, generally straight midvein. Specimens ranging in age from Triassic to Cretaceous have been assigned to this genus, and it is a common form in the Triassic Chinle Formation.

Figure 69. *Cladophlebis*

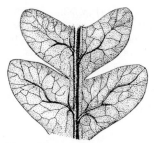

Figure 70. *Cynepteris*

Cynepteris

Cynepteris (Pl. 27, fig. 1) has a twice-pinnate frond with oblong pinnules that are connected at their base (Fig. 70). Lateral veins form a netlike (reticulate) pattern. That the lateral veins do not connect with one another near the pinnule margins is characteristic of *Cynepteris.* Isolated sporangia are scattered over the lower surface of the fertile pinnules of this genus from the Chinle Formation.

Schizaeaceae

Lygodium (climbing fern)

The living representatives of *Lygodium* are true climbers. They twine two or three feet high on saplings along watercourses from New England to Florida. Attached to these twining stems are short-stalked, heart-shaped (cordate), generally palmately lobed segments (Fig. 71). Fossil specimens of this genus have been reported from several Tertiary floras, such as those of the Green River (Pl. 8, fig. 2), and the Chalk Bluff.

Gleicheniaceae

Gleichenia

Figure 71. *Lygodium* (climbing fern)

Living species of *Gleichenia* are widely distributed throughout the tropics of both hemispheres, subtropical eastern Asia, and the humid regions of the southern zone. The fossil species are equally numerous and widespread. They occur in various floras throughout the Cretaceous, particularly the Dakota Formation of Utah and the Smithers area of Canada, and became extinct in this region during the Tertiary. By Pleistocene time, *Gleichenia* occurred in the West Indies, South America, Southeast Asia, and Africa, where it presently grows. Leaves are usually small and are from one to three times pinnate. Often, in rather complete species, the unique equal division of the branching of the leaf at the apex of the petiole can be observed (Fig. 72; Pl. 26, fig. 5). Broadly attached pinnules arise obtusely from the rachis. These are ovate to deltoid and

Figure 72. *Gleichenia* frond

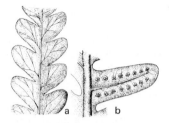

Figure 73. *Gleichenia.* a. Close-up of sterile pinnules, b. Enlargement of fertile pinnule, showing the sori

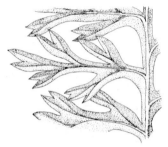

Figure 74. *Wingatea*

Figure 75. *Coniopteris.* a. Sterile foliage, b. Enlargement of modified fertile pinnules with sporangia on tips

have obtuse to acute apices (Fig. 73a). The midvein is distinct, and there are many secondary veins each of which divides, producing from two to five veinlets before reaching the pinnule margins. Sori, when present, are nonindusiate and round, and they occur on both sides of the midvein as rings of four to eight globose sporangia (Fig. 73b).

Wingatea

Wingatea (Pl. 6, figs. 4 and 5) has been reported from only two localities in the Chinle Formation. It has tripinnate fronds. The secondary pinnae are divided into small and delicate pinnules with apices (tips) that may be rounded, pointed, or deeply cleft into two or three lobes (Fig. 74). A single vein enters each pinnule but may branch one or more times, each vein ending a short distance from the pinnule margins. The upper base of the pinnule is deeply indented or constricted, almost to the midrib, whereas the lower portion of the pinnule is decurrent and joins the next pinnule below. The fertile pinnules are similar to the sterile ones but have scattered sori.

Dicksoniaceae

Coniopteris

Coniopteris (Pl. 7, fig. 4) is probably the best-known fossil genus of the Dicksoniaceae. Fronds are bipinnate or tripinnate, with the sterile portion near the base and the fertile portion near the tip. In different species the sterile pinnules are lanceolate, rhomboidal, or elliptical (Fig. 75a). Margins are entire on pinnules near the pinnae apex and become dissected toward the pinnae base. Apices of the pinnules are acute, and their attachment is basal. Veins are strongly ascending with one vein per pinnule, and they branch to provide a veinlet for each lobe. Fertile pinnules are reduced, alternate, and connected along the rachis (Fig. 75b). The end of each reduced lobe contains a subcircular sorus that, in turn, is enclosed in an indusium. This genus is known from the Jurassic and Cretaceous, particularly the Dakota Formation.

Figure 76. *Phlebopteris*

Figure 77. *Matonidium*. a. View of branching, b. Fertile pinnule

Figure 78. *Clathropteris*

Matoniaceae

Phlebopteris

Phlebopteris and *Matonidium* are characterized by the division of the frond into two equal branches (Fig. 76). Each of these branches is subdivided into a series of pinnae (Pl. 27, fig. 5). Each pinnule (leaflet) has a broad basal attachment, a distinct midvein with many smaller lateral veins, and a tapering or bluntly rounded tip or apex. The sori are in a row on each side of the midvein on a fertile pinnule (see *Matonidium*). In *Phlebopteris,* the sori are without an indusium. *Phlebopteris* has been collected from the Triassic Chinle Formation.

Matonidium

Matonidium (Pl. 27, fig. 4) is distinguished from *Phlebopteris* by the presence of an indusium (Fig. 77b). *Matonidium* commonly occurs in Jurassic and Cretaceous rather than Triassic floras. It is common in some western localities of the Cretaceous Dakota Formation.

Dipteridaceae

Clathropteris

The fronds of *Clathropteris* superficially resemble certain maple leaves and appear round in overview (Pl. 27, fig. 2). They are of moderate size, ranging from about 1½ inches (3 cm) in young branches to nearly 10 inches (25 cm) in mature ones. The leaf consists of three to five pinnae (Fig. 78), each of which is linear-lanceolate with toothed margins. These pinnae are united into a basal web. The fertile pinnae have sori irregularly distributed on their lower surface. This genus has been collected from the Chinle Formation.

65

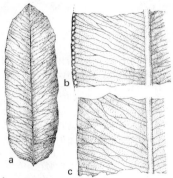

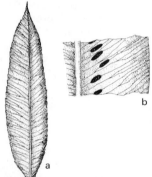

Figure 79. *Saccoloma gardneri.* a. Single pinnule, b. Portion of fertile pinnule, showing a marginal band, c. Close-up of pinnule, illustrating venation pattern

Figure 80. *Allantodiopsis erosa.* a. Single pinnule, b. Sori and venation

Figure 81. *Asplenium.* a. Fertile pinnules, b. Portion of pinna

Polypodiaceae

Saccoloma gardneri

The venation of the lanceolate pinnules of *Saccoloma gardneri* is characteristic of this species (Fig. 79a). It has a conspicuous midvein that extends to the apex. The numerous laterals that diverge from the midvein begin by ascending sharply upward and then curve abruptly outward to the margins.

The fertile pinnules of *Saccoloma gardneri* generally have along their margins a carbonized band thought to be the remains of an indusium or an infolding of the margin over a sporangial band, forming a false indusium (Fig. 79b).

Saccoloma gardneri is very similar in appearance to *Allantodiopsis erosa* (Fig. 80a). They are often found associated with each other in Cretaceous and Paleocene floras. *Saccoloma gardneri* has secondary or lateral veins that often fuse to make a netlike pattern (anastomose) (Fig. 79c), whereas the veins in *Allantodiopsis erosa* never do (Fig. 80b). Also, the pinnae of *A. erosa* are generally smaller.

Asplenium (spleenwort fern)

The pinnules of *Asplenium* vary considerably as to shape and size. The only consistent feature of these ferns is the sori, which are seldom preserved in the fossil record (Pl. 26, fig. 2). The sori of this genus are straight or slightly curved. Their indusium is attached by one edge on the outer side of the fertile vein and is open toward the midvein (Fig. 81a). The name *Asplenium* is from the Greek and applies to the supposed medicinal properties of the plant as a curative for spleen diseases. Fossil forms have been collected from Cretaceous and Tertiary strata.

Astralopteris

Fronds of *Astralopteris* consist of large, long pinnules attached either alternately or oppositely on the rachis (Pl. 26, fig. 1). The pinnules are attached by a single point near the lower portion of the frond (Fig. 82b) and by their complete base on the upper part (Fig. 82a). Major secondary veins arise acutely from a distinct midvein and eventually become perpendicular to it. The sori occur in a single row on each side of the midvein between the major secondary veins (Fig.

66

82b). Tertiary veinlets form a network pattern between the secondaries.

Astralopteris has been reported from the Dakota Formation in Utah and northern Arizona and from the Frontier Formation in Wyoming.

Dryopteris (shield fern)

This fern is common in the Florissant flora (Pl. 6, fig. 6) as well as other Cretaceous and Tertiary floras. The fronds are twice pinnate (Fig. 83a). The pinnules are crenate or lobed, and those that are fertile have roundish sori or fruit dots (Fig. 83b). Secondary or lateral veins divide twice after arising from the midvein.

Figure 82. *Astralopteris.* a. Apex of pinna, b. Fertile pinnules

Figure 83. *Dryopteris* (shield fern). a. Pinna apex, b. Fertile pinnules

Fernlike Foliage

Figure 84. *Crossopteris.* a. Portion of pinna, illustrating lobing, b. Enlarged portion of pinnule, showing venation

Figure 85. *Neuropteris.* a. Portion of pinna, demonstrating lobing of pinnules and pinnule attachment, b. Pinnule

The term *fernlike foliage* is applied to foliage of a fernlike appearance, of unknown affinities, and usually of Carboniferous or Permian age. Originally, the fernlike leaves of these ages were thought to have belonged entirely to ferns, which reproduce by spores. However, near the turn of the century, it was shown that many of these leaves were assignable to the pteridosperms or seed ferns, which have foliage similar to ferns but which reproduce by seeds. (See Ferns for a discussion of the morphology of the frond or leaf.)

The plants bearing fernlike foliage probably included herbaceous forms having a general fernlike structure. Some were slender-stemmed scramblers, relying for support on the surrounding vegetation, whereas others may have had the habit or growth structure of tree ferns, growing to 2 feet or more in thickness and 70 feet in height.

Neuropteridaceae

Crossopteris

Crossopteris (Pl. 25, fig. 2) represents a fernlike foliage type from the Manning Canyon Shale whose true affinities are presently unknown. It is characterized by having pinnules attached along the rachis by one-half of their bases and by undulating margins that eventually give way to lobes downward on the pinnae (Fig. 84). The venation and pinnule shape of this genus are similar to those of *Neuropteris,* but *Neuropteris* has a single-point attachment and forms lobes at the bases of its larger pinnules (Fig. 85).

68

Odontopteris

Odontopteris (Fig. 86) pinnules are small, decurrent, and often wedge shaped. They lack a definite midvein (Pl. 7, fig. 2); instead, the veins enter the pinnule directly from the rachis and fork several times, producing secondary veins that arch before meeting the margins.

Sphenopteridaceae

Sphenopteris

Sphenopteris (Fig. 87) is a common Carboniferous genus into which are placed sterile leaves of species having similar pinnule shape, attachment, and venation. However, *Sphenopteris* is a provisional group from which both ferns and seed ferns have been separated. Pinnules of *Sphenopteris* are variable in shape (Pl. 7, fig. 5; Pl. 22, fig. 6). Generally they are small, wedge shaped, rounded, more or less constricted at the base, and divided into angular or rounded lobes (Fig. 88). Each pinnule has a single midvein that divides, providing each lobe with one or two veins.

Rhodea

Rhodea (Pl. 22, fig. 3; Pl. 26, fig. 4) is similar to some species of *Sphenopteris* but has more delicate fronds with deeply divided, linear pinnules. This Carboniferous genus may also be distinguished from *Sphenopteris* by the single midvein essentially filling the pinnule.

Figure 86. *Odontopteris* pinnule

Figure 87. *Sphenopteris* sp.

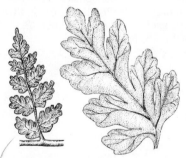

Figure 88. *Sphenopteris* sp.

Zeilleria

Zeilleria (Fig. 89) has *Rhodea*-like pinnules with fructifications attached at the pinnule tips (Pl. 6, fig. 1). At maturity these fruiting bodies may split into four or five valves or segments. This genus occurs in the Manning Canyon Shale.

Sphenopteridium

Sphenopteridium (Fig. 90) is distinguished by short, subtrilobed to trilobed pinnules. Pinnules are wedge shaped with truncated or rounded apices (Pl. 25, fig. 1). The veins divide and parallel the pinnule margins before striking the borders. One of the species of this genus *(S. dissectum)* from the Manning Canyon Shale flora is considered Mississippian in age.

Figure 89. *Zeilleria*

Pecopteridaceae

Pecopteris

Pinnules of this genus are attached to the rachis by the complete width of their base (Fig. 91). These pinnules are generally short, with entire and parallel margins. The midvein in *Pecopteris* normally extends to the apex of the pinnule, and secondary or lateral veins divide once (rarely twice) and are often grouped together. *Pecopteris* has been collected from the Honaker Trail Formation near Moab, Utah (Pl. 25, fig. 6), and from the Spotted Ridge Formation of Oregon *(P. oregonensis)* (Pl. 25, fig. 5).

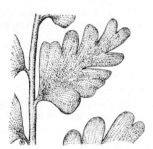

Figure 90. *Sphenopteridium*

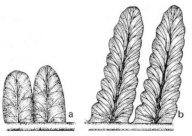

Figure 91. *Pecopteris* pinnules. a. *Senftenbergia* (*Pecopteris*) *pennaeformis*, b. *Pecopteris lamurensis*

70

Seeds and Microsporangiate Structures

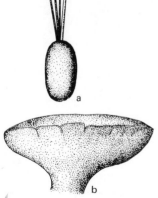

Figure 92. Seeds. a. *Gnetopsis,*
b. *Carpolithus radiatus,*

Seeds are varied and generally isolated, that is, not attached to any foliage; consequently, they are treated as separate form genera.

Cardiocarpus

Cardiocarpus (Fig. 92d, f) is the genus for heart-shaped (cordate), compressed seeds from Pennsylvanian or Permian floras (Pl. 4, fig. 6 and Pl. 5, fig. 1). A narrow border or wing forms the outer portion of these seeds, which originally may have been attached to either cordaitean or seed-fern types.

Trigonocarpus

Trigonocarpus (Fig. 92e) is a common seed form from the Manning Canyon Shale flora. Seeds assigned to this genus are larger and more oblong than those of *Cardiocarpus*. These radially symmetrical seeds are characterized by three longitudinal ribs, which may or may not be present in compressions (Pl. 5, fig. 2). Generally one or two of these ribs can be observed. Similar seeds have been reported attached to *Neuropteris* foliage.

Gnetopsis

Gnetopsis (Fig. 92a) represents small seeds of unknown affinities with four slender appendages extending from the micropylar end (Pl. 4, fig. 5). They are abundant in some collecting localities in the Manning Canyon Shale flora of Utah.

71

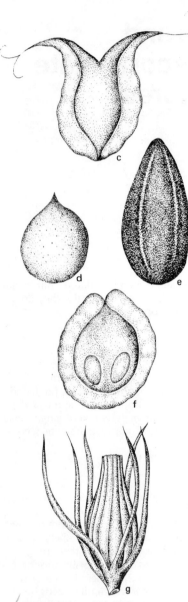

Cornucarpus longicaudatus

Cornucarpus longicaudatus (Fig. 92c) is a small seed from the Manning Canyon Shale with long, narrow whiplike projections that extend from near its micropyle (Pl. 4, fig. 3), divide, and curve back. Although this form is not abundant, it is readily identifiable, as is *Rigbyocarpus* (Fig. 92g), which consists of bracts surrounding a jug-shaped seed.

Aulacotheca

Aulacotheca (Fig. 93) also from the Manning Canyon Shale, is an elongated, cylindrical or narrowing, club-shaped seedlike pollen- or spore-bearing body marked with longitudinal furrows or ridges (Pl. 4, fig. 2).

Figure 92 (cont'd.). c. *Cornucarpus longicaudatus,* d. *Cardiocarpus jayshuleri,* e. *Trigonocarpus,* f. *Cardiocarpus binutus,* g. *Rigbyocarpus*

Carpolithus

Various fruit and seedlike fossils that cannot be definitely placed have been grouped into the catchall genus *Carpolithus* (Fig. 94). Seedlike casts collected from the Morrison Formation have also been placed in *Carpolithus* (Fig. 92b). A few of these (Fig. 96), however, have cell structure and appear similar to *Pityocladus,* the genus for short shoots of a conifer that grew during Morrison time (p. 93).

Ficus

Some seedlike casts from the Upper Cretaceous Foxhill Sandstone in western North Dakota have been assigned to *Palmocarpon* (palmlike seeds). Others from this formation, although similar to *Equisetum* bulbils, have been described as *Ficus,* that is, as possible fig fruits (Fig. 95).

Figure 93. *Aulacotheca* with tip removed to show interior of microsporangiate structure

Figure 94. *Carpolithus vitaceous*

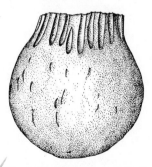

Figure 95. *Ficus ceratops* (fig?) fruit

73

Figure 96. *Carpolithus provoensis*

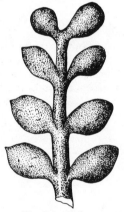

Figure 97. *Behuninia* reconstruction

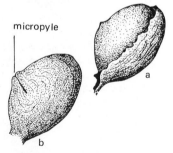

micropyle

Figure 98. *Jensensispermum*. a. Seed with pulp removed, b. Seed with portion of integument or endosperm present

Behuninia

Seeds assignable to *Behuninia* are common in the Jurassic Morrison Formation. The exact relationship of these seeds to any group in the plant kingdom is presently unknown. However, they are similar to some seeds of the Cycadales and are opposite to subopposite on a short stalk or sporophyll (Fig. 97). The younger seeds tend to be smaller and nearly rounded, whereas older seeds are more elongated and ovate (egg shaped) or pyriform (pear shaped).

Jensensispermum

Other berrylike seeds associated with *Behuninia* are called *Jensensispermum.* They are nearly ovate and have a pulpy integument of which only a portion may remain (Fig. 98). *Jensensispermum* resemble seeds of living cycads.

Cycadophyta
(Cycads)

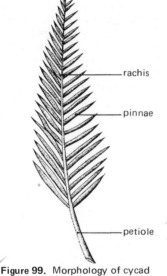

Figure 99. Morphology of cycad frond

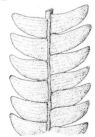

Figure 100. *Nilssonia*

This group is composed of two orders: Cycadales and Cycadeoidales. The former refers to the living cycads (Pl. 38, figs. 1 and 2); the latter includes most of the fossil forms. Living cycads are tropical or subtropical plants with large fern-like leaves (Fig. 99) forming a crown or rosette around the top of the stem. Leaves of the two orders are similar, and often microscopic cuticle studies are required to differentiate them. Cycads are palmlike gymnosperms with cones as fruiting structures. They are not, however, related to palms, which are angiosperms or flowering plants.

Cycadales (or Nilssoniales)

Nilssonia

To *Nilssonia* are assigned leaves of plants that are related to the true cycads and that often occur as members of Mesozoic floras. The linear to lance-olate leaves are compound, having a prominent rachis and, if preserved, a petiole. The lamina of each pinna is attached to the upper surface of the rachis (Fig. 100). Simple veins arise nearly at right angles to the rachis and curve toward the outer margin.

Ctenis

Ctenis is a large, pinnate leaf composed of pinnae attached alternately and suboppositely to the

75

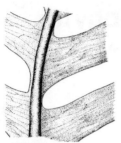

Figure 101. *Ctenis*

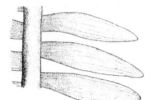

Figure 102. *Pseudoctenis*

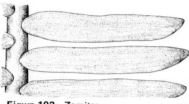

Figure 103. *Zamites*

rachis (Fig. 101). Pinnae are broad and sub-oblong in outline. They taper gradually beyond their middle to an obtuse apex and are attached at right angles to the rachis. These decurrent pinnae are fairly close together. Prominent veins are parallel or subparallel. They divide freely near their point of origin and, occasionally, throughout the pinnae. The veins join together to form a network (anastomose), particularly near their basal origin. These leaves occur in Cretaceous floras of the western United States.

Pseudoctenis

The pinnate leaf of *Pseudoctenis* (Pl. 1, fig. 3) is large (more than 30 cm long) and commonly asymmetric. Pinnae are shorter near the leaf base and gradually increase in length upward in the leaf. Generally they are attached at right angles to the rachis, but in some species they curve downward (Fig. 102). They are attached by their entire base and may be constricted along their upper margins. Pinnae are closely spaced and are widest at or below their middle, tapering to an acute or narrowly rounded apex. Veins parallel the pinnae. A few divisions of the veins occur near their origin and towards the pinnae apex.

This genus has been reported from Lower Cretaceous strata in the West.

Figure 104. Frond of *Zamites*

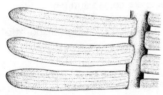

Figure 105. *Ptilophyllum*

Figure 106. Frond of *Ptilo-phyllum*

Cycadeoidales

Zamites

Zamites is a widespread genus for cycadeoid leaves. These are often pinnate with the pinnae attached at right angles to the rachis by a slight, yet distinct, constricted base (Figs. 103 and 104). The overall compound leaf is lanceolate with a single, terminal pinna.

This genus resembles *Ptilophyllum* in the shape of its pinnae (linear or lanceolate), its attachment to the upper surface of the rachis, and its parallel venation. The major difference between *Zamites* and *Ptilophyllum* is in the base of each pinna: that of *Zamites* is more constricted. In the western United States *Zamites* occur commonly in Triassic strata.

Ptilophyllum

Pinnate leaves of this genus are petiolated. Pinnae are generally at right angles to the rachis (Figs. 105 and 106). They are attached alternately or suboppositely to the upper surface of the rachis by their complete bases. Pinnae are linear, straight, often falcate (sickle shaped) and have an acutely or bluntly pointed apex (Pl. 28, fig. 3). A few parallel veins are present. *Ptilophyllum* has been collected from Upper Jurassic and Lower Cretaceous formations of western North America.

Pseudocycas

Fronds of *Pseudocycas* reach 25 cm or more in length. Its pinnae are closely spaced and are commonly inclined at 40 degrees to the rachis (Pl. 1, fig. 2). The pinnae are linear, parallel sided, often contracting acuminately near the apex and terminating in a sharp spine (Fig. 107). Pinnae are 2 to 3 cm in width. Each pinna contains a single midvein and no lateral veins.

Pseudocycas has been reported from western floras of Jurassic and Cretaceous ages.

Pterophyllum

Petiolated, pinnate leaves of this Mesozoic genus have elongated, rectangular-elliptical blades truncated at both ends. Pinnae are linear with nearly parallel margins and are attached by the entire width of their bases (Fig. 108) at right angles to the rachis. Pinnae higher on the rachis may be more oblique. They may vary as to length and width, depending upon their position on the rachis. Veins are coarse and generally from three to six in number. These occasionally fork near the apex but rarely near their point of origin.

Figure 107. *Pseudocycas*

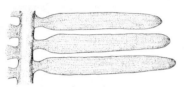

Figure 108. *Pterophyllum*

Figure 109. Frond of *Otozamites lucerensis*

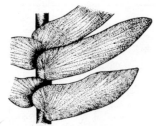

Figure 110. *Otozamites brevi-folius*

Figure 111. *Otozamites lucerensis*

Otozamites

Specimens of this genus have been reported from the Triassic to the Lower Cretaceous (Pl. 28, fig.1).

The frond of *Otozamites* is pinnate (Fig. 109), with alternate pinnae that are either separate or close together or, often, touching (Figs. 110 and 111). Each pinna is attached to the upper portion of the rachis by its base (Figs. 110 and 111). The pinnae are long and narrow, although in some species they may be oval. Their apices are acute or obtuse and have a distinctive auriculate (ear-shaped), asymmetrical base (Fig. 111). Veins usually radiate from the base of the pinnae. However, in more linear forms they have a tendency to be parallel. In both forms they branch occasionally.

Chart 4—Key to Some Cycadeoidean Foliage (courtesy of Sidney Ash)

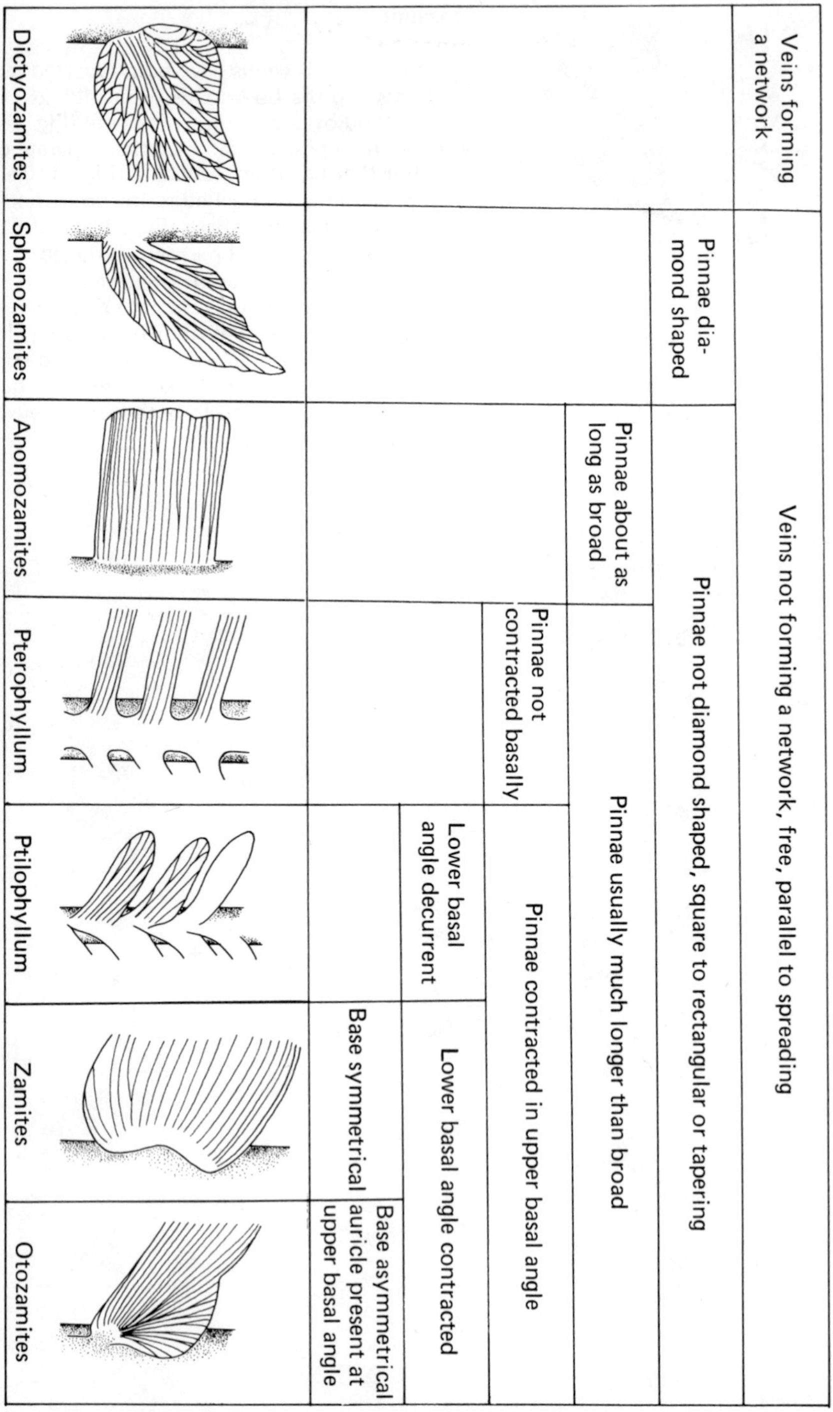

							Genus
Veins forming a network							Dictyozamites
Veins not forming a network, free, parallel to spreading	Pinnae diamond shaped						Sphenozamites
	Pinnae not diamond shaped, square to rectangular or tapering	Pinnae about as long as broad					Anomozamites
		Pinnae usually much longer than broad	Pinnae not contracted basally				Pterophyllum
			Pinnae contracted in upper basal angle	Lower basal angle decurrent			Ptilophyllum
				Lower basal angle contracted	Base symmetrical		Zamites
					Base asymmetrical auricle present at upper basal angle		Otozamites

Ginkgophyta (∼ales)
(Ginkgos)

Figure 112. Living *Ginkgo* (maidenhair fern)

Figure 113. *Ginkgo biloba,* showing leaves and fruits

The development of ginkgos parallels in time the development of the cycads. Their origin was possibly the late Paleozoic. They became widespread by the Jurassic period and then declined during the Tertiary. Ginkgos have been reported in the fossil record of the western United States until the Miocene epoch, when they finally disappeared from this region. *Ginkgo biloba* is the only living species of this group (Fig. 112).

Knowledge of fossil ginkgos is based largely on leaf remains, fossilized fruiting bodies and stems being rarely preserved. Their leaves are fan shaped (Fig. 113). Some leaves are deeply lobed; others are not. The venation of the leaves appears to be parallel, although each vein is divided.

√Coniferophyta

Figure 114. Living *Picea* (spruce)

Figure 115. Female cone of *Pinus* (pine)

Coniferales or conifers, the most familiar group in temperate areas, includes pines, spruces, hemlocks, cedars, and others commonly called evergreens. All are woody types.

Conifers are the most common gymnospermous plants. Some of the living forms, such as the pines, are widely distributed; others, such as the *Sequoiadendron* (giant redwood), are greatly restricted. The growth habit of the conifers produces an appearance distinct from those of other tree-forms. A central trunk with horizontal branches diminishing in length upwards creates a conical outline commonly observed in the spruces (Fig. 114).

The needlelike leaf is another distinctive feature. It is able to endure cold winters because of the small amount of exposed surface area and the tough protective cells that cover the leaf. The leaves do not fall with any regularity, and therefore the trees are referred to as evergreens. Some exceptions are *Larix* (larch), *Metasequoia* (dawn redwood), and *Taxodium* (bald cypress), which lose their leaves each year and therefore are actually deciduous.

Two types of cones are borne on conifers: the large, conspicuous, commonly observed, seed-bearing female cone (Fig. 115) and the small, pollen-bearing male cone (Fig. 116). Both cone types, as well as conifer seeds, are found fossilized.

The Coniferophyta consist of three orders: the extinct Cordaitales, the Coniferales (or conifers), and the Taxales. The latter two orders contain both fossil and living genera and species.

82

Figure 116. Male cone of *Pinus* (pine)

Figure 117. Reconstruction of *Cordaites*

Cordaitales

Cordaitales were large trees that often attained heights of 100 feet (Fig. 117). These trees, distantly related to living conifers such as pines and firs, were widely distributed during Pennsylvanian and Permian times.

The trunk of the genus *Cordaites* was rather slender and unbranched except at the crown, which had many scattered branches. The trunk had a thin-walled and partially hollow pith in the center, which, like the pith in *Calamites*, formed a characteristic cast. Known generically as *Artisia* (Fig. 118; Pl. 29, fig. 2), these casts were formed in a manner unlike that characterizing the formation of *Calamites* casts. The young stems of *Cordaites* had a large and solid pith, which, because it did not keep pace with the elongation of the growing stem, ruptured at intervals, leaving an occasional diaphragm of tissue (Fig. 118b). Eventually these diaphragms disintegrated and left a more or less continuous hollow cylinder. When the plant died, the hollow pith filled with sediment and a cast was formed. Lines or grooves on the cast depict the remnants of the diaphragms.

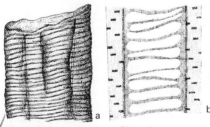

a   b

Figure 118. a. *Artisia,* b. Pith of *Cordaites,* showing diaphragms of tissue

Figure 119. Close-up of *Cordaites,* showing leaves and *Cordaianthus* cones

The leaves of *Cordaites* (Fig. 119), long and straplike, with parallel veins (Pl. 28, fig. 4), are very different from the needlelike leaves of pines.

The fruiting structures assigned to *Cordaites* are loosely compacted cones known as *Cordaianthus* (Fig. 120) that consist of short shoots with tufts of sterile scales from which protruded an elongated sporophyll or stalk bearing small, heart-shaped, pendulous seeds. When isolated, these seeds are assigned to *Cardiocarpus* (Pl. 4, fig. 6 and Pl. 5, fig. 1).

Coniferales

Although *Pseudotsuga* (Douglas fir), *Larix* (larch), and *Sequoiadendron* (giant redwood) are among the many conifers occurring in fossil floras, they are not common enough to be considered here.

— sporophyll
— seed
— sterile scales
— bract

Figure 120. Enlarged portion of *Cordaianthus*

Figure 121. *Walchia*

Lebachiaceae

Walchia. The family Lebachiaceae, to which *Walchia* belongs, has been reported from late Carboniferous and early Permian deposits. Members of this group are known only from the Northern Hemisphere: North America, Europe, and Asia. *Walchia* (Pl. 29, fig. 3) has been collected from the Permian Hermit Shale flora in Grand Canyon.

In general, *Walchia* resembles the modern conifer, *Araucaria heterophylla* (Norfolk Island pine). The main branches of both are borne in whorls of five or six on the trunk (Fig. 121). The secondary branches are two-ranked along the main branches. The needle- or scalelike leaves are spirally arranged on these branches.

84

Araucariaceae

Figure 122. *Araucaria* (Norfolk Island pine)

Araucaria (Norfolk Island pine). Branches and branchlets are thick. Needlelike leaves are lanceolate, recurved, and obtusely pointed (Fig. 122; Pl. 29, fig. 4). They are very close together. The base of the needle is expanded and decurrent. When the needle falls, a deep, obovate, grooved scar remains on the stem.   ~egg-shaped

   Cones (Pl. 9, fig. 5; Pl. 39, fig. 1) composed of rhombic scales (Fig. 123) have been reported attached to *Araucaria*-like foliage. Both the leaves and the cones have been reported frequently in Upper Cretaceous floras in Wyoming, Utah, and British Columbia (?). The cone scales have also been collected separately in Cretaceous floras of eastern Wyoming (Fig. 124).

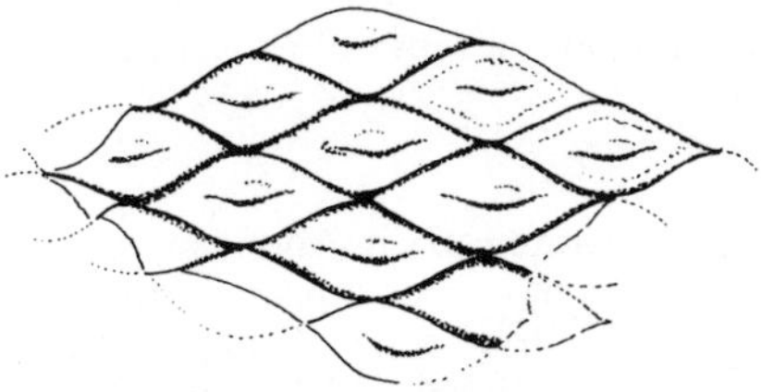

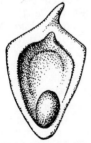

Figure 123. Portion of *Araucaria* (Norfolk Island pine) cone

Figure 124. *Araucaria* (Norfolk Island pine) cone scale

Pagiophyllum. *Pagiophyllum* (Pl. 9, fig. 4) is a genus of conifers supposedly having araucarian affinities. This genus has been reported from Triassic, Jurassic, and Cretaceous floras of western Canada and the United States.

The branching leaf shoots comprising this genus have spirally arranged leaves crowded upon the branches. The needlelike leaves are acutely pointed, decurrent, often slightly falcate (sickle-shaped) and triangular in outline (Fig. 125).

Pending adequate documentation such as the microscopic study of the leaf cuticle, many sterile *Araucaria*-like shoots are referred to *Pagiophyllum.*

Pagiophyllum differs from the similar genus *Brachyphyllum* in that the free part of the leaf of the latter is always short, being no greater in length than the width of its base. The free portion of the leaf of *Pagiophyllum* is longer, and its length is more than the width of its leaf base.

Figure 125. *Pagiophyllum*

Cupressaceae

Chamaecyparis (white cedar). Abundant specimens of this genus have been reported from the Florissant flora and from other Oligocene and Miocene localities in Oregon, Idaho, and Montana. They consist of branchlets with small, scalelike leaves decussately (oppositely) attached (Fig. 128). These ovate, bluntly pointed leaves attached to the branchlets often form flat, horizontal, feathery sprays (Pl. 12, fig. 1).

Figure 126. *Juniper*

In the absence of their cones, *Chamaecyparis, Thuja,* and *Juniper* (Fig. 126) are difficult to distinguish. *Thuja* (Fig. 127), however, may be distinguished by its differing needle arrangement and larger size. In *Thuja* the middle needle overlaps the juncture of the lateral leaves and the scale leaf above; in *Chamaecyparis* the middle leaf does not. The *Chamaecyparis* cones are globose and consist of six to eight decussate, peltate scales (Fig. 129; Pl. 12, figs. 2 and 3).

Figure 127. *Thuja*

Brachyphyllum. This genus has been reported in strata ranging from Permian to Cretaceous ages in the western United States. The fossil remains of *Brachyphyllum* usually consist of slender, elongated branches covered with obliquely spiraled, crowded, appressed leaves (Fig. 130). Leaves are commonly deltoid, bluntly pointed, and often relatively smooth.

Figure 128. *Chamaecyparis* (white cedar)

a   b

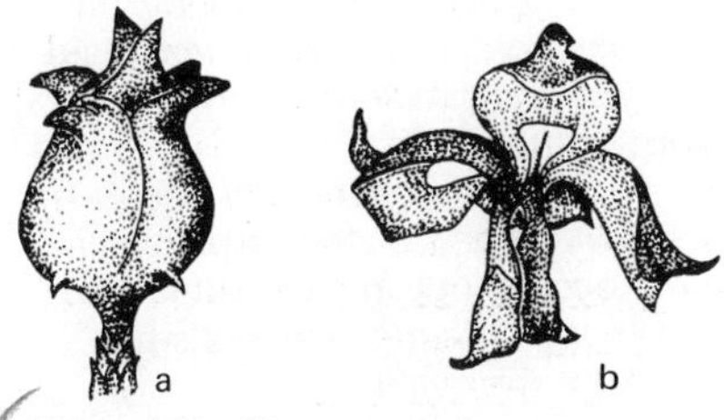

Figure 129. *Chamaecyparis* (white cedar) cones. a. Closed cone, b. Mature expanded cone

Figure 130. *Brachyphyllum*

87

Figure 131. *Sequoia* (redwood) needles

Figure 132. *Sequoia* (redwood). a. Female cone, b. Seed

Figure 133. *Taxodium* (bald cypress) needles

Taxodiaceae

Sequoia (redwood). This genus is represented today by the single species, *Sequoia sempervirens*, which is restricted to a narrow coastal belt in northern California. *Sequoia* has been reported from floras of Upper Cretaceous and Tertiary ages. Leaves of this genus are alternate, flattened and decurrent (Fig. 131) and have acute tips and short petioles. Awl-shaped to needle-shaped leaves occur on the fertile shoots. Branchlets and cones (Pl. 30, fig. 4) with stalks having small, scalelike leaves attached have been reported from floras of Eocene, Oligocene, and Miocene ages.

The cones are ovoid and have spirally arranged, peltate scales (Fig. 132a). Their seeds have two wings as broad as the seed (Fig. 132b).

The dominant taxodiaceous form in a flora is determined by the environmental conditions under which that flora grew. *Taxodium* (bald cypress) is largely indicative of swampy habitats. *Glyptostrobus* (water pine) (Pl. 11, figs. 5 and 6), a closely related genus, generally occurs in well-drained lowland areas, along streams or lakes. *Sequoia* occurs, not in areas subject to flooding, but in well-drained lowland sites, usually terraces in river valleys.

Taxodium (bald cypress). Although bald cypress has not been reported from many Tertiary localities, its foliage and male cones are some of the dominant plant fossils in the Mascall, Oregon; Latah, Washington; Elko, Nevada; and Weaverville, California, floras.

The needlelike leaves of *Taxodium* are alternate and acute, becoming outwardly (distally) narrow (Fig. 133). The base of the needle is narrow and appears sessile, but actually is decurrent and parallels the shoot axis (Pl. 9, fig. 1). The needles are very similar to those of *Sequoia* but differ in that the straight, decurrent leaf bases of *Taxodium* are parallel or lacking, whereas the decurrent bases of *Sequoia* run oblique to the axis because of the twisting nature of the petiole.

Figure 134. *Taxodium* (bald cypress) cone

Figure 135. *Metasequoia* (dawn redwood) needles

Spirally arranged peltate scales make up the subglobose female cone, whereas the male cone consists of six to eight microsporangia (Fig. 134).

Metasequoia (dawn redwood). *Metasequoia* was first known as a fossil from Pliocene deposits in Korea. In 1948, living specimens were discovered in a very restricted area of central China, and now the dawn redwood is commonly cultivated in America. The genus has since been found in floras from Upper Cretaceous to Middle Tertiary ages throughout North America, Greenland, and Asia. One fossil species, *Metasequoia occidentalis,* is a dominant form in floras ranging in age from Paleocene into the Middle Miocene of North America. At that time it grew from northern Alaska to California and eastward into North Dakota.

These fossil remains often consist of foliage shoots with needles decussately (oppositely) attached (Fig. 135; Pl. 30, fig. 2). *Metasequoia* differs from *Sequoia* by the opposite attachment of its needles, the opposite branching of its foliage shoots, and the vertical alignment of the scales on its cones.

Ovoid cones (Pl. 12, fig. 5; Pl. 30, fig. 3) borne on characteristic naked, long stalks are fairly common in many of these floras (Fig. 136).

Figure 136. *Metasequoia* (dawn redwood) cone

Figure 137. *Pinus* (pine) needles

- apophysis
- umbo
- prickle

Figure 138. *Pinus* (pine) cone

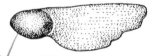

Figure 139. *Pinus* (pine) seed

Figure 140. *Abies* (fir) needles

Pinaceae

Pinus (pine). Leaves, cones, and winged seeds of pine are numerous in Tertiary floras in the western United States. Pine leaves are needlelike, occurring in bundles (fascicles) of one to five and, rarely, six (Fig. 137). The needle number per fascicle varies according to the species of pine. For example, *Pinus florissanti* (Pl. 11, fig. 4), a relative of the living *Pinus ponderosa* (yellow pine), with three long, stout needles and *P. Wheeleri,* with four needles per fascicle, are two rather common fossil species. Because of the fragmentary nature of much of the fossil material, species of *Pinus* are the most difficult to evaluate among the fossil conifers. Fascicles may have needles missing, thus making the true number of needles difficult to determine.

Cones (Pl. 11, fig. 3) are long, cylindrical ovals and consist of spiral rows of hard, woody scales (Fig. 138) each of which may be somewhat thickened on its exposed terminal surface (apophysis). This terminal surface may also have an oblong or diamond-shaped (rhomboidal) extension or scar (umbo). These umbos indicate the previous year's growth. Depending upon the species, small, sharp prickles may be present on the umbos.

Oval pine seeds attached to thin, terminally rounded, papery-appearing wings are often collected (Fig. 139).

Abies (fir). Fir leaves, although usually linear, are sometimes curved, flat, and rather blunt (Pl. 10, fig. 5). The single, sessilely attached, spirally arranged needles generally extend from all sides of the twig (Fig. 140). When the leaves fall, they leave a conspicuous, smooth, rounded scar on the twig.

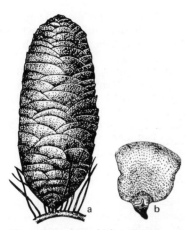

Figure 141. *Abies* (fir). a. Female cone, b. Cone scale

Cones are oblong and cylindrical (Fig. 141a), with cone scales falling at maturity from the cone axis (Pl. 10, fig. 6), which may also be preserved. The small, broad-oval or fan-shaped cone scales are often fossilized (Fig. 141b). These are broadly rounded at the distal edge of the scale and gradually taper to a slender stalk at the point of attachment. A bract may or may not extend beyond the distal scale margin. In some species, the bract may be surmounted by a spiny tip exserted or extended beyond the scale.

Long, oval seeds of *Abies* (Pl. 10, figs. 2-4) are attached to a large, broad wing that is symmetrically rounded above (Fig. 142).

Fossil fir remains are common in Tertiary floras. The closely related fossil genus *Abietites* has been reported from the Upper Cretaceous.

Tsuga (hemlock). Slender leafy branchlets of hemlock illustrating a year's growth have been reported. These bear conspicuous leaf scars and various sizes of needles. Leaves assignable to hemlock are spirally arranged, single, flattened or rounded, abruptly petiolate (Fig. 143), and often attached to the twig by a twisted petiole. Persistent leaf bases (sterigmata) remain when the needles fall.

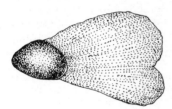

Figure 142. *Abies* (fir) seed

Figure 143. *Tsuga* (hemlock) needles

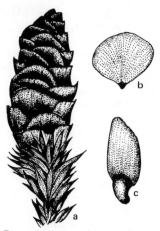

Figure 144. *Tsuga* (hemlock). a. Female cone, b. Cone scale, c. Seed

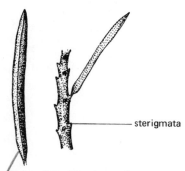

sterigmata

Figure 145. *Picea* (spruce) needles

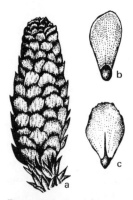

Figure 146. *Picea* (spruce). a. Female cone, b. Seed, c. Cone scale

Cones are oblong to ovoid and consist of nearly orbicular to obovate, entire-margined cone scales (Fig. 144a). These taper to a wedge-shaped (cuneate) base (Fig. 144b).

Seeds are attached to long, usually obovate wings (Fig. 144c). The *Tsuga* seed is widest near the middle of the wing which characteristically extends under the lower edge of the seed. It differs from the seeds of *Pinus* and *Picea* in that the latter are narrower near the middle of the wing and are attached to the wing.

Picea (spruce). Terminal leafy shoots of spruce are slender with narrow, linear, angular, single needles (Fig. 145) attached on all sides of the twig and clustered on the shoots. Needle tips are acute to obtuse. Needles leave persistent basal peglike sterigmata on the twig when they fall (Pl. 12, fig. 6). Spruce cones are oblong-cylindrical and are composed of numerous obovate to ob-cuneate cone scales (Fig. 146a). These scales are often gently rounded above and wedge-shaped (cuneate) below (Fig. 146c). They are much longer than their bracts.

Spruce seed compressions are numerous in some floras of mid-Tertiary age (Pl. 11, figs. 1 and 2). This may be the result of favorable transportation from more distant elevations above their point of decomposition. These seeds are attached to broad, thin, compressed wings (Fig. 146b).

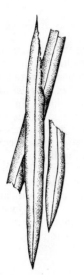

Figure 147. *Pityophyllum*

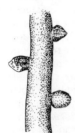

Figure 148. *Pityocladus*

Figure 149. *Protophyllocladus*

Pityophyllum. The generic name *Pityophyllum* is applied to detached needlelike leaves similar to, or broader and flatter than, those of *Pinus* (Pl. 12, fig. 4). Occasionally these are found attached to shoots, but normally they are detached (Fig. 147).

Pityophyllum is abundantly represented in Jurassic and Lower Cretaceous floras. Because they are not assignable to a specific conifer, they presently have little botanical value.

Pityocladus. The branch axes of this genus bear spirally disposed shoots that are 1 to 1.5 cm long. Shoots marked with scars of leaf bases or bracts (Fig. 148) are often found in association with *Pityophyllum* leaves.

Specimens of *Pityocladus* have been reported from Jurassic and Lower Cretaceous floras in the western United States.

Podocarpaceae

Protophyllocladus. Leaflike appendages or phylloclades of this genus are common in many Upper Cretaceous floras. They exhibit a short petiole and generally are oblong-obovate to oblong-lanceolate in outline (Pl. 30, fig. 1). The apex is usually obtuse, rarely pointed. The base is narrowly cuneate (wedge shaped) to the short petiole (Fig. 149). Margins are entire below, becoming obtusely dentate or undulate above. The midvein is distinct throughout most of the phylloclade. Lateral veins are numerous, branched, and generally parallel.

This fossil genus has probable relationship to the modern genus *Phyllocladus*, which grows in New Zealand.

Figure 150. *Cephalotaxus* needles

Figure 151. *Torreya*

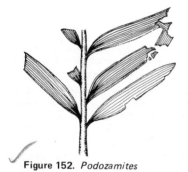

Figure 152. *Podozamites*

Cephalotaxaceae

Comments near a pt.

Cephalotaxus. Leaves are long, lanceolate, slightly curved needles gradually tapering from a slightly broad base to a subacuminate tip (Fig. 150). An exception, *Cephalotaxus nevadensis* Axelrod, is wider at the middle. A short petiole is present, and the midvein is flanked by broad stomatal bands that appear as fine lines.

Cephalotaxus differs from fossil species of *Cunninghamia* in that its needles taper more gradually to a very sharp aristate tip. The *aul-shrr* stomatal bands of *Cephalotaxus* are not in grooves, as is characteristic of the similar *Torreya* (Fig. 151), but are on the needle surface. *Keteleeria* differs from *Cephalotaxus* in having shorter needles with a rounded tip and in not having the broad, stomatal bands.

Coniferales incertae sedis

Podozamites. Shoots of this genus have spirally arranged leaves that are often twisted, giving the appearance of attachment in a single plane. The leaves are lanceolate (Pl. 29, fig. 5) and constricted from their bases to a narrow point of attachment (Fig. 152). The apex may be obtuse or bluntly pointed. Veins are subparallel and crowded. They divide in the basal portion of the leaf and converge slightly near the leaf apex.

Leaves assignable to *Podozamites* were originally thought to be cycadean but are now considered to be related to the conifers, possibly the living genus *Agathis,* which is a member of the araucarian family. *Agathis* currently grows in the Southern Hemisphere.

Detached *Podozamites* leaves are difficult to distinguish from other genera; for example, the pinnae of some species of the cycadeoid *Zamites*.

Anthophyta
(Angiosperms)

Angiosperms or flowering plants are the dominant plant type on earth today. They have become a highly adapted group so numerous and diversified that a distinct characterization of the group is difficult.

The origin of the angiosperms is a mystery. Lower Cretaceous floras contain few or no angiosperms, but by Upper Cretaceous time the angiosperms had spread to every continent. Since many of the earliest angiosperms resemble advanced forms, this group appears to have been in existence long before the Cretaceous period. Most fossil remains of this group are fruits, seeds, and vegetative remains such as leaves and petrified stems.

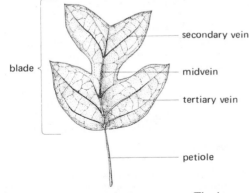

blade

secondary vein

midvein

tertiary vein

petiole

Figure 153. *Liriodendron* (tulip tree)

The leaves of angiosperms are their most conspicuous part. If a leaf is examined closely, a broad, thin blade marked into small divisions by veins can be observed (Fig. 153). The prominent central vein is called the midvein. In some leaf forms, there are several prominent veins known as principal or primary veins. Arising from the primaries are other veins referred to as lateral veins or secondaries, which, in turn, may give rise to tertiary veins. The blade is generally connected to the stem or branch by a petiole.

Four chief types of vein arrangements are present in angiosperm leaves (Fig. 326): (1) parallel—veins lie parallel to one another throughout the leaf, (2) net or reticulate—veinlets branch and then reunite (anastomosing) to produce a fine net or "chicken-wire" effect, (3) pinnate—a prominent midvein and several secondary veins essentially parallel each other to the leaf margins, and (4) palmate—several veins radiating from the base of the midvein like the fingers of a hand. *Acer* (maple) and *Pelargonium* (geranium) leaves are palmately veined.

Angiosperm leaves vary greatly in shape, but usually the leaves of the same genus are recognizably similar in type. Shapes can be classified as linear, lanceolate, ovate, and orbicular or as some modification of these such as oblong, obovate, oval, cordate (heart-shaped) (Fig. 329).

Linear leaves are long and narrow, as illustrated in corn or grasses in general. A lanceolate form is broader just above the base and tapers to a slender tip as in a spear point. An ovate leaf has a base that is broader than the base of a lanceolate leaf, and the outline of an orbicular leaf is nearly circular.

Leaf margins are important in identifying fossil angiosperm leaves (Fig. 330). In some leaf forms the margins are smooth or entire. Some leaves have a series of small notches or teeth. Such margins are referred to as undulate, serrate, dentate, crenate, or lobed. Deep indentations between veins in the leaves of plants (e.g., maple) are called lobing. The notch between each pair of lobes, called the sinus, may be shallow in some forms (oak) and deeply indented in others.

Dicotyledonae

Magnoliaceae most primitive

Magnolia Leaves are simple, entire margined, and unlobed; with either acute or acuminate apices (Fig. 154). They may be oblong, lanceolate (Pl. 33, fig. 5), elliptical, or nearly oval. The midvein is thick near the base of the leaf and becomes

Figure 154. *Magnolia*

Figure 155. *Cercidiphyllum articum*

Figure 156. *Cercidiphyllum* sp. (katsura tree)

Figure 157. *Sassafras*

Figure 158. *Cinnamomum*

thinner toward the apex. Secondaries consist of 4 to 15 pairs of alternate to subopposite veins curving toward the margins, often joining the veins above. Tertiaries and smaller veinlets are at right angles to the secondaries. *Magnolia* has been reported from the Cretaceous to the Recent.

Cercidiphyllaceae

Cercidiphyllum (katsura tree). Leaves of *Cercidiphyllum* are variable. Most have a typical ovate-elliptic (Pl. 19, figs. 1 and 2), sometimes deltoid shape with a rounded base and toothed margins. Depending upon the species, the margins may also be entire. Five veins are palmately arranged, one being a strong midvein at the center (Figs. 155 and 156). Two pairs of alternate secondaries arise from the midvein along its upper part. Secondaries form loops near the margins, and veinlets are numerous.

Cercidiphyllum is widespread and abundant in the Upper Cretaceous and Paleocene. Fruits, seeds and leaves of this genus are often found associated.

Lauraceae

Sassafras. Leaves are oval to obovate and are single, double, or triple lobed (Fig. 157; Pl. 19, fig. 4). They display considerable variation in size but have similar venation. Primary veins diverge from near the base of the leaf and are flanked by veins that loop upward toward the margins. Secondary veins alternate with strong tertiary veins, forming large, irregular polygons, *Sassafras* occurs from Cretaceous to Recent.

Cinnamomum. Leaves of *Cinnamomum* are ovate to elliptic in shape with entire margins and an acute apex (Fig. 158). The base of the blade narrows at the petiole. This form is characterized by a primary vein dividing into five veins near the base of the blade. Two of these veins are intermediate between the marginal veins and the midvein. Specimens assigned to this genus are reported from Cretaceous and Tertiary floras.

Figure 159. Living *Mahonia* (Oregon grape)

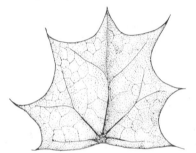

Figure 160. *Mahonia* (Oregon grape) leaflet

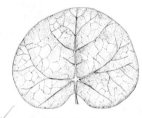

Figure 161. *Menispermites*

Berberidaceae

Mahonia (Oregon grape). *Mahonia* is a common shrub in the present forests of the Sierra-Nevada, Cascade, Coast, and Rocky Mountain ranges (Fig. 159). Some modern species are almost identical to those in the Miocene floras of the western United States. *Mahonia* has also been reported from strata of Eocene and Oligocene ages.

Mahonia leaves are compound, consisting of oblong leaflets with a retuse apex and often an asymmetrical, wedge-shaped (cuneate) base (Pl. 13, fig. 6). Margins may be entire, but generally they are composed of prominent spinose teeth (Fig. 160). The midvein is often strong and straight, giving rise to numerous pinnately arranged secondaries. Secondaries and intersecondary veinlets loop within the margins to form a complex mahonoid venation. Coarse tertiary veins join to form elongated polygons.

Menispermaceae (moonseed family)

Menispermites. Leaves may or may not be palmately lobed. They are elliptical to deltoid-ovate in general outline (Fig. 161; Pl. 19, fig. 3; Pl. 32, fig. 1). Depending upon the species, lobes are bluntly pointed to rounded. Some, however, may be acute or acuminate. The base is cordate (heart shaped) and is attached with a petiole. Margins are entire, and venation is palmate.

Simaroubaceae

Ailanthus (tree of heaven). *Ailanthus* fruits are winged (Fig. 162). They have an oval seed that is placed centrally in the samara but nearer one margin than the other (Pl. 20). The samara is long and ovate, with linear, reticulate venation. The living species of *Ailanthus* is native to the mountains of China. It was brought to California by the Chinese miners and is a familiar shade tree. The fossil remains are common in the Sucker Creek, Trout Creek, Stinking Water, and Florissant floras.

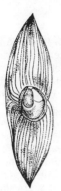

Figure 162. *Ailanthus* (tree of heaven) fruit

Rutaceae

Ptelea (hop tree). Orbicular winged fruits are the usual fossilized remains of this genus (Fig. 163b; Pl. 18, fig. 3). The wings are net veined and emarginate (indented at their bases and apices). The fruits are borne on a slender petiole that is often preserved. The tapering fusiform or ovate seed cavity is divided nearly equally by a vertical groove.

These occur as fossils in the Florissant, Mascall, Sucker Creek, Trout Creek, Stinking Water, and Latah formations. The living species of *Ptelea* grow along dry, rocky slopes on open streamside areas from southern New England westward to southern Colorado and southward to Florida and Texas.

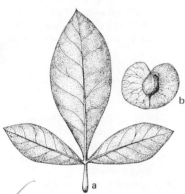

b

a

Figure 163. *Ptelea* (hop tree). a. Leaf, b. Fruit

These fruits are similar to the samara fruit of *Ulmus* (elm). The fruits differ in that—

1. the midvein of elm fruits is on one side of the wing, giving the wing an asymmetrical appearance, and is more central in *Ptelea,*
2. *Ptelea* fruits are borne on a short petiole which is lacking on elm fruits,
3. veins on the surface of the seed cavity are vertical in elms and horizontal in *Ptelea,* and
4. the apex of the *Ptelea* wing is incised or indented.

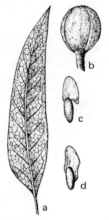

Figure 164. *Cedrela*. a. Leaflet, b. Seed capsule, c-d. Seeds

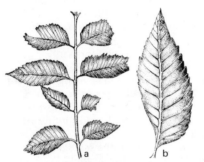

Figure 165. *Astronium*. a. Portion of leaf, b. Leaflet

Figure 166. Calyx of *Astronium* flower

Meliaceae

Cedrela. The genus *Cedrela* has been uncovered in floras of Eocene, Oligocene, and Miocene ages in California, Colorado, Idaho, Oregon, Utah, and Washington. These include the Florissant, Chalk Bluffs, Green River, Latah, and Sucker Creek floras.

Leaves of this genus are pinnately compound. Their leaflets are ovate to lanceolate. The asymmetry of the lateral leaflets (Fig. 164a) is caused partly by the frequent arching of the midvein, which deflects the tip backward slightly, producing leaflets that are somewhat curved and quite oblique. Apices of the leaflets are slenderly acuminate, and they have entire or crenate-serrate margins. The leaflets are attached with slender petiolules. The midvein of the leaflets is curved and stout, and secondary veins are irregularly spaced and subparallel. Twelve to 20 pairs of secondaries alternate with other conspicuously short secondaries. The secondary veins curve upward and loop marginally.

Oval seed capsules (Fig. 164b) having three valves separated by longitudinal sutures have been collected from the Sucker Creek flora. Their seeds are winged and resemble maplelike samaras. Seeds are ovate and their wings are membranous (Fig. 164c, d). Some seeds show a spinous extension (Pl. 13, fig. 4).

Living species of *Cedrela* are widespread in the tropics of Mexico, South America, the West Indies, Southeast Asia, the East Indies, and Australia. These are commonly large trees, and many are used for lumber. The so-called Mexican cedar, *C. mexicana*, is used in manufacturing cigar boxes.

Anacardiaceae

Astronium. Leaves are even-pinnately compound (Fig. 165a). Leaflets are unequally ovate with acute apices, are petiolulated, and have asymmetric bases (Fig. 165b). Margins may be serrate or crenate-dentate. The midvein is strong and gives rise to 12 to 14 opposite to nearly opposite secondaries. The tertiary veins are reticulate.

Calyces of flowers having five sepals are preserved (Fig. 166). The sepals are oblong and have occasionally branched, subparallel veins (Pl. 17, fig. 5). Living species of the genus are presently growing in the American tropics.

100

Figure 167. *Rhus* (sumac). a. *R. stellariaefolia*, L. *R. lesquereuxi*

Figure 168. Leaflet of *Sapindus* (soapberry)

Specimens of *Astronium* have been collected from the Florissant, Green River, and Beaverhead floras.

Rhus (sumac). Leaves assignable to *Rhus* have been collected from various Tertiary formations throughout the West. These include the Florissant, Chalk Bluffs, and Green River floras.

Leaves are compound with ovate-lanceolate leaflets. Leaflets are unequal in symmetry, and their bases are rounded to narrowly wedge shaped (cuneate) (Fig. 167b). Often they are entire near their base, with their apices narrowing abruptly to a sharply pointed tip. The leaflets are either shortly petioluled or attached by their bases (sessile) in some species. On some species upper margins are dentate (Pl. 15, fig. 1; Pl. 28, fig. 5); in others they are entire (Fig. 167a). Eight to 10 secondary veins arise oppositely from the midvein. These do not loop, and the tertiary veins are reticulate.

Some living species of *Rhus,* such as *R. glabra,* are known as the common sumac of the East and Southeast. *R. copallina* var. *lanceolata* is frequently encountered along seasonally dry streams in Texas, New Mexico, and Mexico, while *R. toxicodendron radicans* is the common poison ivy.

Sapindus (soapberry). Leaves of this genus are pinnately compound with 6 to 9—sometimes as many as 20—pairs of subopposite to opposite leaflets (Pl. 18, fig. 4). The leaflets are unequal and rhombic-lanceolate (Fig. 168). They have an acuminate apex, and the upper portion of the base is rounded, whereas the lower portion of the base may be partially heart shaped (cordate). They are attached with a short petiolule. The margins are entire. There are 8 to 10 pairs of irregularly spaced secondaries that arise from a stout midvein, curve upward, and loop with those above. Tertiary venation is reticulate. Living species of *Sapindus* occur in the southern states. Specimens of this genus are common in Tertiary floras and are frequently encountered in the Florissant of Colorado.

Figure 169. *Koelreuteria* (golden-rain tree). a. Portion of leaf, b. Leaflet

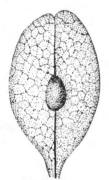

Figure 170. Fruit of *Koelreuteria* (goldenrain tree)

Koelreuteria (goldenrain tree). Leaves of *Koelreuteria* are unevenly pinnate to bipinnate and are composed of several pairs of completely dissected leaflets. These leaflets are ovate and have acute or acuminate apices. Their bases are wedge shaped (cuneate), and their margins are doubly serrate (Fig. 169). Secondary veins arise from a strong midvein and extend into the marginal dentations.

Fruits of *Koelreuteria* are inflated, ovate, leaflike capsules with rounded apices and wedge-shaped bases (Fig. 170; Pl. 15, fig. 4). Midveins are not symmetrical in the fruits.

Specimens of this genus have been uncovered in the Florissant fossil beds.

Cardiospermum (balloon vines). Living members of *Cardiospermum* are herbaceous or woody vines and are common in the West Indies. One species, *C. halicacubum,* presently grows from Missouri into Mexico. The fossil species appear to represent shrubs rather than vines.

Cardiospermum leaves (Pl. 14, fig. 1) are compound and generally occur in segments of threes, although some may have more (Fig. 171). These vary from three-parted basal divisions to deeply divided, decurrent lobes or sometimes even a single lobe. Terminal portions of the leaves commonly have a pair of opposite lobes and, rarely, three to four.

The midvein of these leaves is slender, producing two to three pairs of thin, undivided secondary veins. The stronger secondaries enter the apices of the lobes, whereas other secondaries may loop along the margins. The mesh of the tertiary venation consists of thin veinlets.

Specimens of *Cardiospermum* are common in the Green River and Florissant floras.

Figure 171. *Cardiospermum* (balloon vine) leaflets

Figure 172. *Acer* (maple) leaf

Acer (maple). This genus has an extensive geologic past, dating from the Cretaceous. Leaves are petioled, simple or compound (Fig. 173). Simple leaves are palmately (3 to 7) lobed, often deeply three lobed (Fig. 172; Pl. 31, fig. 1). Leaves are truncate at the base. Venation is strong, and, depending upon the number of lobes, three veins (three-lobed species) or five veins (five-lobed species) divide from the petiole near the base of the blade. Lateral veins pass up and fork around the sinus with three pairs of nonlooping (camptodromic) secondaries below the first small lobes. Depending upon the species, the margins may be either entire or toothed.

The fruit is a samara consisting of a winged seed (Fig. 174). Wings are often wider at their distal end (Pl. 31, fig. 2), tapering gradually to the large obovate seed.

Figure 173. *Acer fremontensis*

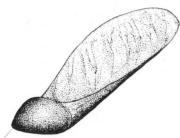

Figure 174. *Acer* (maple) samara

Figure 175. *Ulmus* (elm) leaf

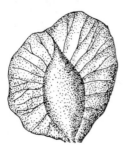

Figure 176. *Ulmus* (elm) fruit

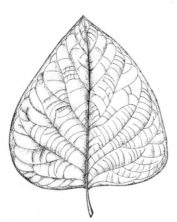

Figure 177. *Ficus alabamensis*

Ulmaceae

Ulmus (elm). Leaves of this genus are lanceolate to obovate in shape. They are petiolate, generally unequal at their bases (Fig. 175), and have acuminate tips. Margins are simply or, more commonly, doubly serrated. The midvein is usually strong and straight with secondary veins arising in 18 or more alternate to opposite pairs. Secondaries are close, parallel, and slightly curved upward, and they end in the marginal teeth. Some secondary veins may branch (not a tertiary venation) before reaching the margins. Finer venation is often not preserved.

Fruits are obovate to oval, often hairy, samaras (Fig. 176). The seed cavity may be indistinct, and the fruit apex may be entire or slightly notched.

Moraceae

Ficus. The genus *Ficus* has been used as a "catchall" taxa for species that with more information could probably be assigned to another genus. Leaves classified with *Ficus* have been reported from floras of Cretaceous and Tertiary ages.

Fossil leaves of this genus have been described as either large (Fig. 177) or small (Fig. 178), obovate, lanceolate, broadly ovate, or nearly circular in outline (Fig. 177). Their bases may vary from rounded to truncate to obtusely wedge shaped (cuneate), and their apices may vary from rounded to obtuse, sometimes acuminate. Leaves are petiolate to sessile and have entire margins.

Figure 178. *Ficus* sp. (fig?)

Figure 179. Venation of *Ficus* (fig?)

Leaves having palmate or pinnate venation have been described. In the pinnately veined forms (Pl. 19, fig. 5), the midvein is strong with irregular or regular ascending, nonlooping (camptodromic) secondaries arising from it (Fig. 179). Smaller cross veins may be present.

Leguminosae

Mimosites. The small, entire, narrowly obovate leaflets of the compound leaves of this fossil genus are widest in their lowest fourth (Fig. 180). The leaflets tend to be sickle shaped (falcate) (Pl. 16, fig. 6), with obtusely pointed apices and wedge-shaped (cuneate) bases. Two to 20 or more pairs of closely spaced secondaries arise from the slender midvein and strike the margins obliquely. Tertiary veins form a complex series of loops resulting in a tertiary netlike (reticulate) pattern.

Fossil leaflets of *Mimosites* indicate that the plant on which they grew may have been adapted to subhumid or even arid conditions. Specimens of this genus are common in the Green River flora.

Figure 180. Leaflets of *Mimosites*

105

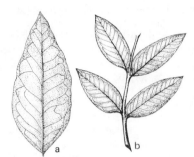

Figure 181. *Caesalpinites acumi-natus.* a. Leaflet, b. Portion of compound leaf

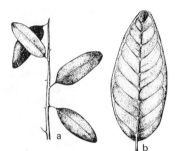

Figure 182. *Robinia* (locust). a. Portion of leaf, b. Leaflet

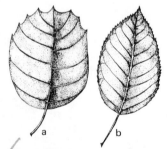

Figure 183. *Amelanchier* (service-berry). a. Fossil, b. Living

Caesalpinites. Leaves of *Caesalpinites* are compound, having ovate to obovate leaflets attached to the leaf rachis by their asymmetrical or wedge-shaped (cuneate) bases. The leaflet apices vary from acute (Fig. 181) to emarginate (Pl. 15, fig. 3). The midvein is slender and will give rise to 3 to 11 pairs of secondary veins. Depending upon the species, these pairs will be either alternate or opposite. The secondaries often branch outward and loop along the leaflet margins. Tertiary venation is reticulate.

The leaflets of *Caesalpinites* are common in the Green River flora.

Robinia (locust). Leaves of *Robinia* are pinnately compound. The leaflets are opposite and are ovate to ovate-rhombic, having a rounded base and a rounded apex (Fig. 182), and are attached with a long petiolule. The margins of the leaflets are entire. The midvein is strong and gives rise to 8 to 10 pairs of opposite or subopposite secondaries that are subparallel and generally branched near the margins. They loop upwards and join near the margins (anastomose). Tertiary venation is reticulate.

Legume. These seed pods (Pl. 13, figs. 2 and 3) have been reported from the Florissant flora and, under the genus *Parkinsonia*, from the Beaverhead flora.

Rosaceae

Amelanchier (serviceberry). Leaves are alternate, simple, and rhombic to oval (Fig. 183). Leaf margins are entire, toothed, or finely serrated, with an acute apex or apices. The leaves are attached by slender petioles. Numerous secondary veins arise from the midvein. Subparallel to each other, they loop upward into the entire or toothed margins (Pl. 13, fig. 1).

Figure 184. Living *Crataegus* (black hawthorn)

Figure 185. Fossil *Crataegus* (hawthorn)

a   b

Figure 186. *Cercocarpus* (mountain mahogany). a. Seed, b. Leaf

Crataegus (hawthorn). Leaves of hawthorn are alternate, simple, and long-elliptic in outline. They have serrate, often more or less lobed margins (Figs. 184 and 185). Their apices are blunt to acute, and their bases vary from acute to obtuse. The leaves are attached by a slender petiole. Prominent secondary veins diverge from the midvein into the lobes, and each secondary has several prominent arching tertiaries that supply the margins. The tertiary mesh is irregular.

Cercocarpus (mountain mahogany). The leaves of *Cercocarpus* are oblanceolate, with a somewhat rounded acute apex (Pl. 14, figs. 4 and 5). Their bases are cuneate, and their margins are slightly serrate in the upper half (Fig. 186b). The petiole is stout. The slender midvein gives rise to six to nine opposite pairs of straight secondary veins that may curve slightly into the marginal teeth. Tertiary venation is finely netlike (reticulate).

Cercocarpus leaves are similar to those assigned to *Fagopsis* but differ in having a cuneate base and more spinose marginal teeth. The leaf lamina in *Fagopsis* is not decurrent, and its teeth are more rounded.

Figure 187. *Platanus* (sycamore)

Figure 188. *Platanophyllum*

a   b

Figure 189. *Lomatia*. a. Leaf, b. Fruit

Platanaceae

Platanus (planetree or sycamore). Leaves assignable to *Platanus* are alternate, simple, and boadly ovate (Pl. 19, fig. 6). They have petioles and vary in having three to five lobes (Fig. 187). The margins of the lobes may or may not be coarsely toothed. Subprimary veins of about equal prominence arise from the primary vein more or less above the petiolar attachment of the blade. These rarely show a palmate division as close to the base of the leaf as do those of maple. The veins end in the pointed apices of the lobes. Secondary veins are thin, numerous, straight or slightly curved, and undivided. The veins enter the teeth of leaves with toothed margins. Tertiary veins may form an irregular network.

Platanophyllum. This genus is abundant in the Clarno, Yellowstone, and Chalk Bluff floras. Leaves of *Platanophyllum* are large and nearly equidimensional. They are fan shaped with five to nine symmetrical, projecting lobes separated by shallow V-shaped sinuses (Fig. 188). A short, stout petiole is attached to a broad, wedge-shaped (cuneate) base. Three primary veins flare upward, and the lateral primaries branch one to three times. Numerous secondaries are closely spaced near the base of the leaf. These curve upward and form a network with adjacent secondaries. Finer tertiary veins are reticulate (net-like).

In the Chalk Bluff flora, *Platanophyllum* have been collected matted together. This, coupled with their association with *Persea* and *Platanus*, suggests a possible streamside habitat.

Proteaceae

Lomatia. Leaves of *Lomatia* are pinnately compound (Fig. 189a). The leaflets are alternate and decurrent upon the rachis (Pl. 14, fig. 2). They are entire or may consist of one to three irregular lobes. The leaflets are obovate-lanceolate and have an acute apex. The slender midvein gives rise to one or two pairs of secondary veins. The fruits of *Lomatia* are winged (Pl. 14, fig. 3). The seeds are obliquely oval and have a wrinkled surface (Fig. 189b). A prominent spine or projection extends from the distal end above the samara attachment. The samara is ovate and has four prominent veins that radiate from the seed. These curve upward and eventually join at the

108

apex. *Lomatia* has been collected from the Green River, Beaverhead, and Florissant floras.

Figure 190. *Oreopanax*

Araliaceae

Oreopanax. Leaves are large, cordate-ovoid, and palmately or, possibly, digitately compound (Fig. 190). Often the primary lobes are further divided into secondary lobes. The venation is composed of palmate primaries, opposite secondaries and tertiaries. These leaves have been reported from the Florissant, Sucker Creek, and Green River floras. One living species of this genus lives in the Andean forest of South America at an elevation of 7,200 to 11,400 feet. Here the rainfall is abundant throughout the year with a few short dry periods.

Salicaceae

Populus (poplar). Specimens assigned to this genus are abundant in floras of Upper Cretaceous and Tertiary ages (Pl. 32, fig. 4). Leaves are petioled and are ovate and ovate-lanceolate (Fig. 191) to elliptical in outline with an acute, acuminate to sometimes broadly rounded apex (Fig. 192). The base of the leaf is broadly rounded to wedge shaped (cuneate). Margins are entire or crenate to serrate. The strong midvein has opposite to alternate pairs of strong, nonlooping secondary veins that diverge at an acute angle and are somewhat crowded toward the midvein base. Tertiary veinlets are thin.

Figure 191. *Populus balsamifera*

Figure 192. *Populus latior*

Figure 193. *Salix* (willow)

Figure 194. *Betula* (birch)

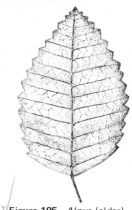

Figure 195. *Alnus* (alder)

Salix (willow). Leaves of this genus are common in Cretaceous and Tertiary floras. They are simple, alternate, and commonly lanceolate (Fig. 193; Pl. 18, figs. 5 and 6). Margins are entire or toothed. They are pinnately veined, sessile, or short petioled. The apex is acuminate in many species. The midvein is strong, with fine, numerous, generally alternate secondaries arising from it. These secondaries curve upward but do not form loops.

Betulaceae

Betula (birch). Leaves are alternate, simple, petioled, and mostly ovate to triangular in outline, often with serrate (Fig. 194; Pl. 33, fig. 1) to lobate margins. Depending upon the species, the base of the leaf may be truncate or, frequently, wedge shaped (cuneate) or heart shaped (cordate), and the apices may vary from acute to acuminate. Its midvein is distinct and has 8 to 12 pairs of subopposite to alternate secondary veins arising from it. Secondaries are rather straight and parallel, ending in the larger teeth, and may have 2 or 3 smaller branches entering the smaller teeth.

Thin, irregular tertiary veins are oblique in the secondaries, and smaller veinlets produce an intricate, fine network between them. Leaves have been reported from many Tertiary floras. To distinguish birch leaves from those of *Alnus* (alders), note that the tips of birch are acuminate to acute and that those of the alder are acute to blunt. The first basal pair of secondaries in birch leaves generally occurs well within the blades—as opposed to the near marginal position of these veins in alder.

Alnus (alder). The *Alnus* leaf is simple and ovate to broadly elliptical, having a truncate to slightly heart-shaped (cordate) base (Fig. 195). Leaves are attached with a thick petiole. Leaf margins are usually biserrate or dentate. The midvein is prominent, with secondary veins alternating and entering the marginal teeth directly. Tertiary veins are netlike (reticulate).

Figure 196. *Alnus* (alder) fruits

Alnus fruits are woody and are ovoid to oblong (Fig. 196). They are composed of truncate scales with thickened tips. Nutlets are nearly circular, having an enclosing membranous wing or two lateral wings.

Specimens assignable to this genus have been collected from the Latah flora in Washington, the Copper Basin flora in Nevada, the Thorn Creek and Payette floras of Idaho, and others (Pl. 15, fig. 5).

Ostrya (hop hornbeam). This genus has been reported from several western Miocene floras but never as an abundant species.

Leaves are simple and oblong-ovate. They have doubly serrate margins with acuminate tips and rounded bases (Fig. 197). The midvein is strong and straight, and between 11 and 15 secondaries arise alternately to oppositely from it. These may divide again near the margins.

Some species of birch may be difficult to separate from species of *Ostrya*, particularly when fragmentary. Birch leaves are generally wider near their bases, and the prominent abaxial tertiaries are rarely as well developed from the secondaries in birch as in *Ostrya*.

Fagaceae

Fagopsis. This extinct genus has been reported from the Florissant, Ruby, Republic (Washington), and Green River floras. Trees supplying leaves in these floras may have been a dominant aspect of the streamside vegetation.

Leaves of this genus are generally ovate with dentate margins, acute apices, and rounded to broadly wedge-shaped (cuneate) bases (Fig. 198). They have a short, strong petiole. The midvein of the leaf is prominent and 10 or 11 pairs of sub-opposite, stout, straight secondary veins arise from it (Pl. 32, fig. 5). The tertiary venation is fine and netlike, although rarely preserved.

Fruits are tiny, ovoid nutlets, and the female cones are similar to those of *Betula* (birch).

Figure 197. *Ostrya*

Figure 198. *Fagopsis*

111

Figure 199. *Carpinus* (hornbeam)

Figure 200. *Quercus* sp. (oak)

Figure 201. *Quercus concinna*

Carpinus (hornbeam). Leaves of this genus are ovate to oblong-ovate (Fig. 199) and have an acuminate apex, a rounded base, and a slender petiole (Pl. 17, fig. 3). Their margins have doubly serrate teeth that extend into slender spines. The midvein is slender and has numerous straight, parallel, commonly opposite secondary veins arising from it. The lower veins are finely divided and enter a second order of teeth.

Carpinus is common in Oligocene and Miocene floras in the West and particularly common in the Florissant flora of Colorado.

Quercus (oak). Leaves of this genus are variable in size and shape (Figs. 200 and 201). They may be entire or characterized by deltoid teeth (Pl. 16, figs. 2-4). The midvein is strong and extends to the apex. The secondary veins are distinct and generally alternate. Secondaries and tertiaries enter the teeth at the margins. Numerous veinlets form a network. Secondaries in some of the entire marginal species arch to join the one above, making an almost continuous line just inside the margins.

Figure 202. *Castanea* (chestnut)

Figure 203. *Dryophyllum*

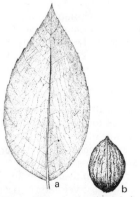

Figure 204. *Juglans* (walnut). a. Leaflet, b. *J. cinerea* (butternut)

Castanea (chestnut). Leaves are simple and oblong-lanceolate, and they have coarsely serrate margins with sharp, large teeth (Fig. 202). The base is wedge shaped (cuneate) and the apex acuminate (Pl. 17, fig. 2). Venation consists of a generally straight midvein giving rise to 13 to 15 pairs of secondaries. Each secondary terminates in a tooth.

These have been reported from Tertiary, particularly Miocene, floras.

Dryophyllum. Leaves of this genus are often oblong-lanceolate (Fig. 203). They have a wedge-shaped (cuneate) base and a gradually narrowing apex or tip (Pl. 17, fig. 4). A petiole should be present but may not be preserved. Margins are serrate. The midvein is strong, with about 15 pairs of regularly spaced, nondividing (craspedo-dromic) secondary veins. These subparallel secondaries curve regularly upward and terminate in the marginal teeth. Tertiary veinlets are thin and are often joined midway between adjacent secondaries by a zigzag vein.

Dryophyllum and *Salix* leaves are similar. However, the secondary veins of *Salix* leaves curve upward (Fig. 193), whereas those of *Dryophyllum* go straight to the margin.

Dryophyllum is one of the dominant forms in Upper Cretaceous floras. Because it is well de-fined, widespread, and abundant, the species *Dryophyllum subfalcatum* Lesq. represents an index fossil of this age. It has been reported from Cretaceous floras in Wyoming and Utah.

Juglandaceae

Juglans (walnut). *Juglans* appears to be one of the oldest fossil dicot genera. It is well repre-sented in floras from Middle Cretaceous to the present. Its leaves are pinnately compound, and its leaflets vary from ovate-lanceolate to oblong-lanceolate (Fig. 204a). They have acuminate to acute apices or drip points (Pl. 18, fig. 1). The bases of the leaflets are often wedge shaped (cuneate). Their margins may be entire or finely serrate. A distinct, straight midvein extends to the apex, and secondary veins are alternate, thin, and upward curving, extending almost to the margins. Veins in some fossil species are non-looping (camptodromous). Finer veins are often not preserved.

113

Figure 205. *Carya* (hickory)

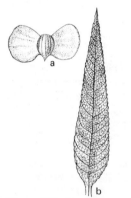

Figure 206. *Pterocarya*. a. Fruit, b. Leaflet

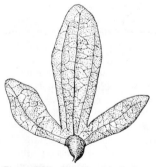

Figure 207. *Engelhardtia* fruit

Carya (hickory). Leaves are pinnately compound with 9 to 11 opposite leaflets that vary in size and shape and have unequal wedge-shaped bases. The leaflets are ovate (Fig. 205) and petiolulated, with acuminate or acute apices. Their margins are strongly serrate. Twelve to 16 pairs of opposite to alternate secondary veins arise from a slender midvein and are nondividing (craspedodromic). Tertiary veinlets are coarse and netlike (reticulate).

The more pronounced dentations on the margins of the *Carya* distinguish them from the leaflets of *Juglans.* Secondary veins are more curved and ascending in *Juglans,* whereas they are straighter in *Carya.*

Leaves of this genus have been reported from the Florissant, Mascall, Latah, and Thorn Creek floras.

Pterocarya (wingnut). Members of this genus are known from Eocene to Recent. They occur in many Tertiary floras, such as the Green River, Florissant, Payette, Sucker Creek, and Latah.

Leaves are compound, and leaflets are petiolate, slender, and lanceolate. The greatest width is in the lower one-third of the leaflet. The base of the leaflet is wedge shaped (cuneate), and the apex is generally acuminate (Fig. 206b). Margins may be evenly dentate. They have a strong midvein from which arise several pairs of parallel secondaries. These curve upward and loop with each succeeding secondary within the margins. Tertiary veinlets form a netlike (reticulate) pattern.

The fruit has opposite wings and resembles a butterfly (Fig. 206a). Longitudinal grooves may be observed on the fruits. Modern members of this genus live along streams in China and India and are planted occasionally in the United States as an ornamental.

Engelhardtia. The fruit of this genus consists of a winged nutlet with three to four wings (Fig. 207; Pl. 16, fig. 5). Margins of the wings are entire. The central wing is longer than the laterals and contains three parallel veins. The shorter primary veins paralleling the midvein are connected to it by secondary cross veinlets. Miocene forms are pinnately veined.

Zelkova (keaki tree). Leaves of *Zelkova* (Pl. 17, fig. 6) are lanceolate and vary in size. Margins are dentate (Fig. 208). The base is wedge shaped (cuneate), and the apex is acute. The midvein is prominent, with distant secondaries opposite, subopposite, or alternate to each other. Each secondary terminates in a tooth. Leaves have been reported from the Bridge Creek, Ruby, and Green River floras.

Monocotyledonae

Typhaceae

Figure 208. *Zelkova* (keaki tree)

Typha (cattail). Leaves are long and linear (Fig. 209). Venation consists of longitudinal, parallel veins with numerous fine cross veins (Pl. 17, fig. 1). *Typha* leaves occur in the Florissant, Mascall, Stinking Water, Latah, Trout Creek, and Sucker Creek floras.

Palmae

Palm leaves have been collected from floras of Upper Cretaceous, Paleocene, and Eocene ages in Utah, Wyoming, Montana, and west Texas. These generally consist of the two genera: *Phoenicites* (Fig. 210), which includes forms previously called *Genomites*; and *Sabalites* (Fig. 211).

Figure 209. *Typha* (cattail)

Phoenicites. *Phoenicites* leaves are very large, measuring up to six feet in length when found complete. The general outline of the leaf is oblong-ovate. The rachis is wider at the base, tapering upward, and rays of the leaf arising from the rachis are united and marked by plications or folds (Pl. 33, fig. 6). The parallel veins on the plications appear as fine lines on the surface.

In separating *Phoenicites* from *Sabalites,* the following criteria generally apply:

1. The rachis is continuous in *Phoenicites* but tapers abruptly to a short rachis in *Sabalites.*
2. The lower rays are acute and have a decurrent divergence in *Phoenicites,* whereas they are spread out fanlike in *Sabalites.*
3. *Phoenicites* has a gradually tapering linear rachis, and *Sabalites* has a rather deltoid rachis; thus, *Phoenicites* is pinnate and *Sabalites* palmate.

Figure 210. *Phoenicites*

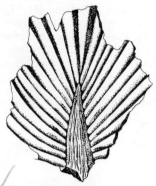

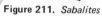

Sabalites. The leaves of *Sabalites* are palmate, with numerous rays arranged like a fan (Pl. 33, fig. 4). A large, stout petiole enlarges into a deltoid rachis at the base of the leaf.

Figure 211. *Sabalites*

Figure 212. Reconstruction of *Sabalites*

Plate 1

Fig. 1 *Podozamites* (Coniferophyta) 1/2X (after Bell, 1956)

Fig. 2 *Pseudocycas* (Cycadophyta) (after Bell, 1956)

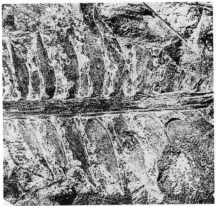

Fig. 3 *Pseudoctenis* (Cycadophyta) (after Bell, 1956)

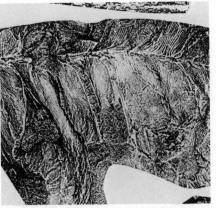

Fig. 4 *Ctenis* (Cycadophyta) 1/2X (after Bell, 1956)

Fig. 5 *Ptilophyllum* (Cycadophyta) (after Bell, 1956)

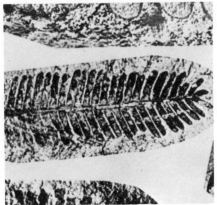

Fig. 6 *Pterophyllum* (Cycadophyta) (after Bell, 1956)

Plate 2

Fig. 1 *Lepidophylloides* (Lycopodophyta)

Fig. 2 *Neocalamites* (Sphenophyta)

Fig. 3 *Equisetum* (horsetail) (Sphenophyta)

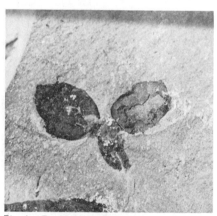

Fig. 4 *Equisetum* bulbils

Fig. 5 *Equisetum:* longitudinal section showing whorls of branches at node

Fig. 6 *Equisetum:* cross section of stem with branches

Plate 3

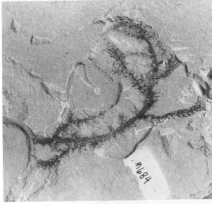

Fig. 1 *Asterophyllites* (Sphenophyta)

Fig. 2 *Annularia* (Sphenophyta)

Fig. 3 *Lepidophloios* (Lycopodophyta)

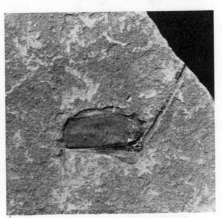

Fig. 4 *Lepidocarpon* (Lycopodophyta)

Plate 4
Dendrites and Seeds

Fig. 1 Dendrites: false fossils

Fig. 2 *Aulacotheca*

Fig. 3 *Cornucarpus longicaudatus*

Fig. 4 *Cornucarpus acutum*

Fig. 5 *Gnetopsis*

Fig. 6 *Cardiocarpus* sp.

Plate 5
Dendrites and Seeds (cont'd.)

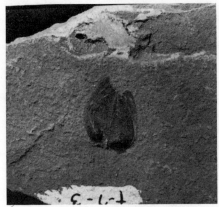
Fig. 1 *Cardiocarpus binutus*

Fig. 2 *Trigonocarpus*

Plate 6

Ferns and Fernlike Foliage

Fig. 1 *Zeilleria* (fernlike foliage)

Fig. 2 *Osmunda* (fern) and *Metasequoia* (conifer) (lower right)

Fig. 3 *Neuropteris* (fernlike foliage)

Fig. 4 *Wingatea* (fern)

Fig. 5 *Wingatea*

Fig. 6 *Dryopteris* (shield fern)

Plate 7
Ferns and Fernlike Foliage (cont'd.)

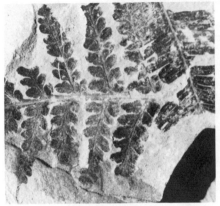

Fig. 1 *Cladophlebis* (fern)

Fig. 2 *Odontopteris* (top) and *Pecopteris* (fernlike foliage)

Fig. 3 *Alloiopteris* (fernlike foliage)

Fig. 4 *Coniopteris* (fern)

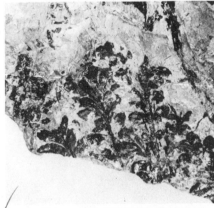

Fig. 5 *Sphenopteris* (fernlike foliage)

Fig. 6 *Asplenium* (spleenwort fern)

Plate 8
Ferns and Fernlike Foliage (cont'd.)

Fig. 1 *Matonidium* (fern) sori with indusia   Fig. 2 *Lygodium* (climbing fern)

Plate 9
Coniferophyta

Fig. 1 *Taxodium* (bald cypress)

Fig. 2 *Thuja*

Fig. 3 *Thuja* branch

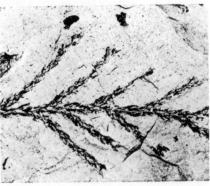

Fig. 4 *Pagiophyllum* (after Bell, 1956)

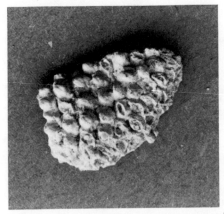

Fig. 5 *Araucaria* (Norfolk Island pine) cone

Fig. 6 *Sequoia* (redwood) cone

Plate 10
Coniferophyta (cont'd.)

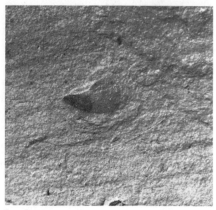

Fig. 1 *Abies* (fir) cone scale with *Abies* needle

Fig. 2 *Abies* seed

Fig. 3 *Abies* seed

Fig. 4 *Abies* seed

Fig. 5 *Abies* needle

Fig. 6 *Abies* cone

Plate 11
Coniferophyta (cont'd.)

Fig. 1 *Picea* (spruce) seed

Fig. 2 *Picea* seed

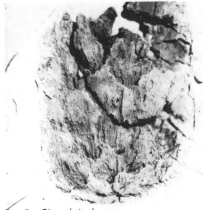
Fig. 3 *Pinus* (pine) cone

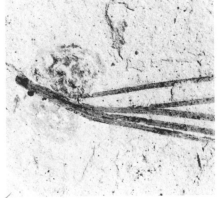
Fig. 4 *Pinus* needles

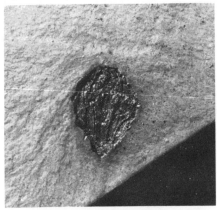
Fig. 5 *Glyptostrobus* (water pine) cone

Fig. 6 *Glyptostrobus* foliage

Plate 12
Coniferophyta (cont'd.)

Fig. 1 *Chamaecyparis* (white cedar) foliage

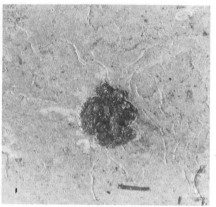

Fig. 2 *Chamaecyparis* slightly expanded cone

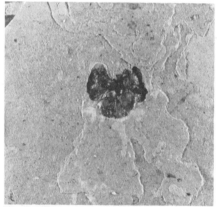

Fig. 3 *Chamaecyparis* expanded cone

Fig. 4 *Pityophyllum* 1/2X (after Bell, 1956)

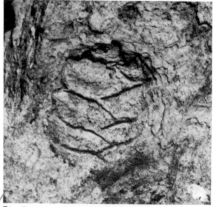

Fig. 5 *Metasequoia* (dawn redwood) cone

Fig. 6 Sterigmata on spruce twig

Plate 13
Angiosperms

Fig. 1 *Amelanchier* (serviceberry)

Fig. 2 Legume

Fig. 3 Legume possibly belonging to
Robinia (locust)

Fig. 4 *Cedrela* seed

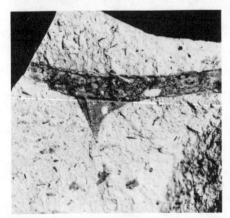

Fig. 5 *Crataegus* (hawthorn) thorn

Fig. 6 *Mahonia* (Oregon grape)

Plate 14
Angiosperms (cont'd.)

Fig. 1 *Cardiospermum* (balloon vine)

Fig. 2 *Lomatia*

Fig. 3 *Lomatia* seed

Fig. 4 *Cercocarpus* (mountain mahogany)

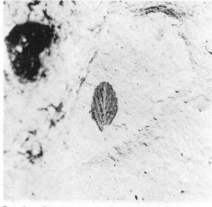

Fig. 5 *Cercocarpus* leaf: partial

Fig. 6 *Rhus* (sumac)

Plate 15
Angiosperms (cont'd.)

Fig. 1 *Rhus* leaflet

Fig. 2 *Rhus stellariaefolia* leaflet

Fig. 3 *Caesalpina* (divi-divi tree)

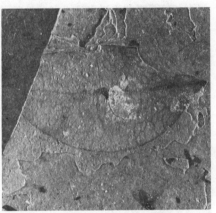

Fig. 4 *Koelreuteria* (goldenrain tree) fruit

Fig. 5 *Alnus* (alder)

Fig. 6 *Alnus* cone

Plate 16
Angiosperms (cont'd.)

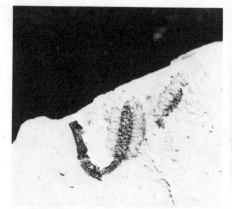

Fig. 1   *Alnus* cone

Fig. 2   *Quercus hannibali* (oak) leaf

Fig. 3   *Quercus* (oak)

Fig. 4   *Quercus:* species with entire margins

Fig. 5   *Engelhardtia* fruit

Fig. 6   *Mimosites* leaflet

Plate 17
Angiosperms (cont'd.)

Fig. 1 *Typha* (cattail)

Fig. 2 *Castanea* (chestnut)

Fig. 3 *Carpinus* (hornbeam)

Fig. 4 *Dryophyllum*

Fig. 5 *Astronium* flower

Fig. 6 *Zelkova* (keaki tree)

Plate 18
Angiosperms (cont'd.)

Fig. 1 *Juglans* (walnut) leaf

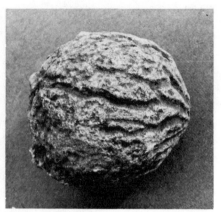

Fig. 2 *Juglans* fruit

Fig. 3 *Ptelea* (hop tree) fruit

Fig. 4 *Sapindus* (soapberry)

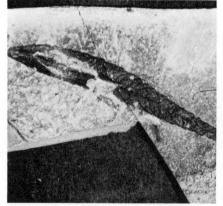

Fig. 5 *Salix* (willow)

Fig. 6 *Salix*

Plate 19
Angiosperms (cont'd.)

Fig. 1   *Cercidiphyllum* (katsura tree)

Fig. 2   *Cercidiphyllum*

Fig. 3   *Menispermites* (after Bell, 1956)

Fig. 4   *Sassafras*

Fig. 5   *Ficus*

Fig. 6   *Platanus* (sycamore)

Plate 20
Angiosperms (cont'd.)

Fig. 1   *Ailanthus* (tree of heaven) seed

Plate 21
Reconstruction of the Oligocene Florissant Flora in central Colorado
(painting by permission of Stuart Heimdal)

Plate 22
Algae and Fernlike Foliage

Fig. 1 Stromatolite

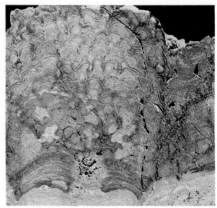

Fig. 2 *Chlorellopsis*

Fig. 3 *Rhodea*

Fig. 4 Algal balls: cross section

Fig. 5 *Collenia*

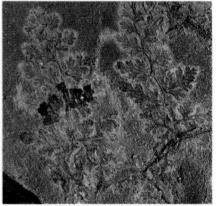

Fig. 6 *Sphenopteris schatzlarensis*

Plate 23
Sphenophyta and Lycopodophyta

Fig. 1 *Mesocalamites*

Fig. 2 *Calamites suckowi*

Fig. 3 *Palaeostachya*

Fig. 4 *Archaeocalamites*

Fig. 5 *Archaeocalamites* foliage

Fig. 6 *Lepidostrobophyllum*

Plate 24
Lycopodophyta

Fig. 1 *Stigmaria*

Fig. 2 *Lepidostrobus ornatus*

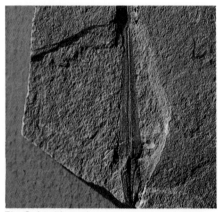

Fig. 3 *Lepidostrobus squarrosus* sporophyll

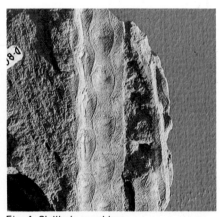

Fig. 4 *Sigillaria canobiana*

Fig. 5. *Sigillaria*

Fig. 6 *Lepidodendron mannabachense*

Plate 25
Fernlike Foliage

Fig. 1 *Sphenopteridium dissectum*

Fig. 2 *Crossopteris utahensis*

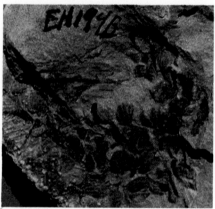

Fig. 3 *Odontopteris* sp.

Fig. 4 *Odontopteris* sp.

Fig. 5 *Pecopteris oregonensis*

Fig. 6 *Pecopteris* sp.

Plate 26
Ferns and Fernlike Foliage

Fig. 1 *Astralopteris coloradica*

Fig. 2 *Asplenium dakotensis*

Fig. 3 *Corynepteris* (*Alloiopteris*) sp.

Fig. 4 *Rhodea* sp.

Fig. 5 *Gleichenia* sp.

Fig. 6 *Sphenopteris* sp.

Plate 27
Ferns

Fig. 1 *Cynepteris*

Fig. 2 *Clathropteris*

Fig. 3 *Cladophlebis*

Fig. 4 *Matonidium*

Fig. 5 *Phlebopteris*

Fig. 6 *Wingatea*

Plate 28

Fig. 1 *Otozamites*

Fig. 2 *Cycadeoidea*

Fig. 3 *Ptilophyllum* pinna

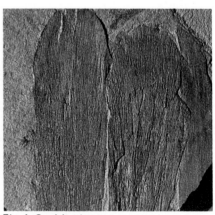

Fig. 4 *Cordaites* leaves

Fig. 5 *Rhus* (sumac) leaves

Fig. 6 *Lagenospermum* cupules

Plate 29
Coniferophyta

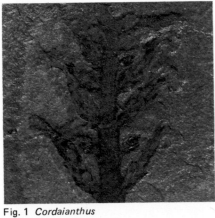

Fig. 1 *Cordaianthus*

Fig. 2 *Artisia*

Fig. 3 *Walchia*

Fig. 4 *Araucaria* (Norfolk Island pine) foliage

Fig. 5 *Podozamites*

Fig. 6 *Schilderia*

Plate 30
Coniferophyta (cont'd.)

Fig. 1 *Protophyllocladus*

Fig. 2 *Metasequoia* (dawn redwood) foliage

Fig. 3 *Metasequoia* cone

Fig. 4 *Sequoia* (redwood) foliage

Fig. 5 *Taxodium* (bald cypress)

Fig. 6 *Abies* (fir)

Plate 31
Angiosperms and Coniferophyta

√Fig. 1 *Acer* (maple) leaf

Fig. 2 *Acer* samara

√Fig. 3 *Platanus* (sycamore)

Fig. 4 *Platanus*

Fig. 5 *Platanus* venation

Fig. 6 *Chamaecyparis* (white cedar)

Plate 32
Angiosperms

Fig. 1 *Menispermites*

Fig. 2 *Fagus* (beech)

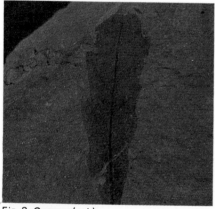

Fig. 3 *Quercus* (oak)

Fig. 4 *Populus* (poplar)

Fig. 5 *Fagopsis*

Fig. 6 *Cardiospermum* (balloon vine)

Plate 33
Angiosperms

Fig. 1 *Betula* (birch)

Fig. 2 *Palmoxylon:* cross section

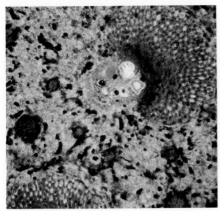

Fig. 3 *Palmoxylon:* cross section of bundle

Fig. 4 *Sabalites*

Fig. 5 *Magnolia*

Fig. 6 *Phoenicites*

Plate 34
Coniferophyta (petrified wood)

Fig. 1 *Araucarioxylon* (petrified log)

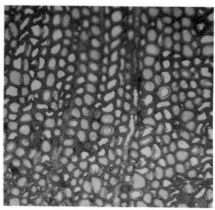

Fig. 2 *Araucarioxylon:* cross section

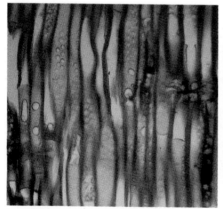

Fig. 3 *Araucarioxylon:* radial pitting

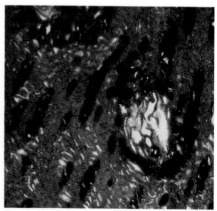

Fig. 4 *Pityoxylon:* radial section. Note vertical resin canals.

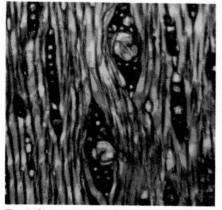

Fig. 5 *Pityoxylon:* tangential section. Note horizontal resin canals.

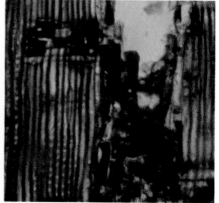

Fig. 6 *Pityoxylon:* close-up of radial section. Note vertical pitting and resin canals.

Plate 35
Xenoxylon

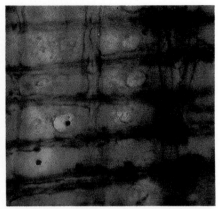

Fig. 1 Tangential section, showing septate tracheids

Fig. 2 Radial section, illustrating crossfield pitting

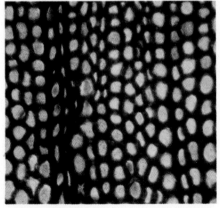

Fig. 3 Cross section

Fig. 4 Close-up of ray and septate tracheid

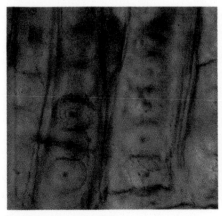

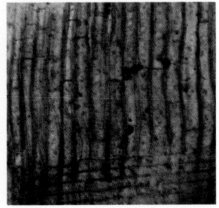

Fig. 5 Close-up of pitting

Fig. 6 Radial section

Plate 36
Paraphyllanthoxylon (petrified wood)

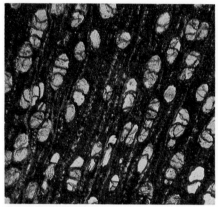

Fig. 1 Cross section

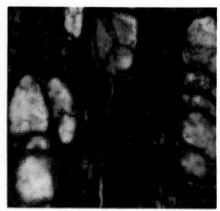

Fig. 2 Cross section, enlarged

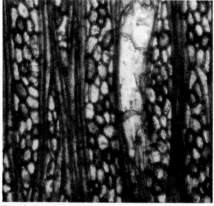

Fig. 3 Tangential section

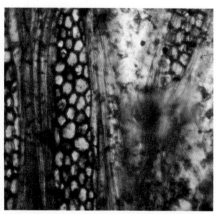

Fig. 4 Tangential section, slightly enlarged

Fig. 5 Radial section

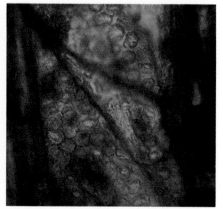

Fig. 6 Close-up of vessel, showing pitting

Plate 37

Fig. 1 *Tempskya* (petrified fern): cross section of false trunk, illustrating smaller roots surrounding larger stems

Fig. 2 *Tempskya:* longitudinal section of base of false trunk, showing branching of stem

Fig. 3 *Tempskya* ¼X: portion of log

Fig. 4 *Osmundacaulis* (petrified fern): rhizome

Fig. 5 *Osmundacaulis* 12X: close-up of rhizome

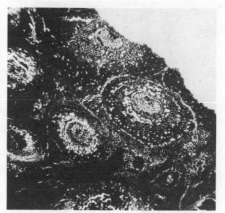

Fig. 6 *Osmundacaulis* 7X: close-up of leaf traces

Plate 38

Fig. 1 *Cycas* (living cycad)

Fig. 2 *Zamia* (living cycad): note cones at apex of stem

Fig. 3 *Rhizopalmoxylon* 5X: cross section of palm root

Fig. 4 *Monanthesia* (cycadeoid) ¼X: cross section

Fig. 5 *Monanthesia* 1/40X: reconstruction

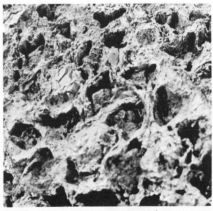

Fig. 6 *Monanthesia:* close-up of stem surface

Plate 39

Fig. 1 *Araucaria* (Norfolk Island pine) 2X: cone

Fig. 2 *Cycadeoidea:* enlarged cone embedded among leaf bases

Fig. 3 *Betula* (birch)

Fig. 4 *Fagara* ¼X

Fig. 5 *Codonophycus* (algae)

Fig. 6 *Woodworthia*

Plate 40
Coniferophyta (cross sections)

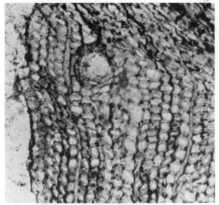

Fig. 1 *Protopiceoxylon*

Fig. 2 *Protopiceoxylon*

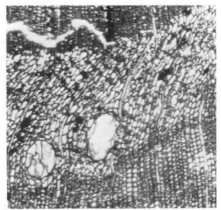

Fig. 3 *Pityoxylon*

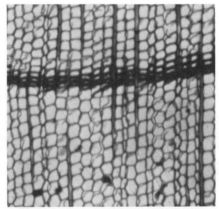

Fig. 4 *Sequoia* (redwood)

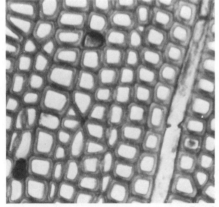

Fig. 5 *Sequoia:* showing resin cells

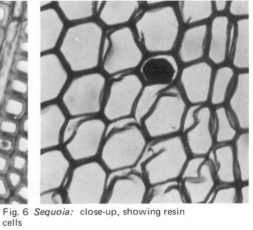

Fig. 6 *Sequoia:* close-up, showing resin cells

Plate 41
Coniferophyta (cross sections) (cont'd).

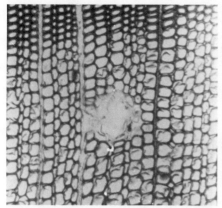

Fig. 1 *Pinus* (pine)

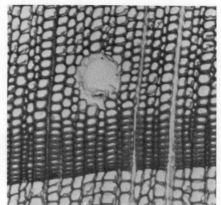

Fig. 2 *Pinus*

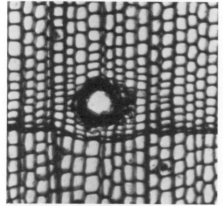

Fig. 3 *Pinus*

Plate 42
Coniferophyta (radial [r] and tangential [t] sections)

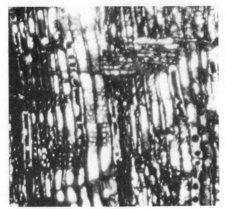

Fig. 1 *Cupressinoxylon* (r): illustrates resin cells and pitting

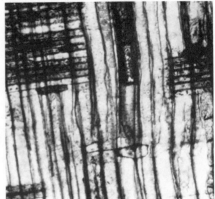

Fig. 2 *Protopiceoxylon* (r)

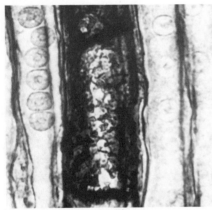

Fig. 3 *Protopiceoxylon* (r): close-up, showing pitting

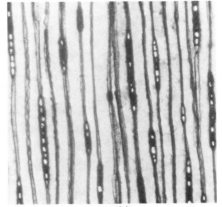

Fig. 4 *Protopiceoxylon* (t)

Fig. 5 *Sequoia* (redwood) (t): showing resin cells

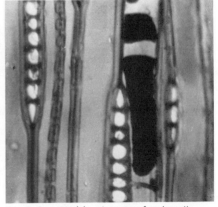

Fig. 6 *Sequoia* (t): close-up of resin cells

Plate 43
Coniferophyta (radial [r] and tangential [t]
sections) (cont'd.)

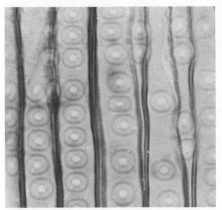

Fig. 1 *Pinus* (pine) (r): close-up, showing pitting

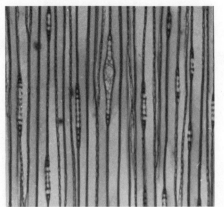

Fig. 2 *Pinus* (t)

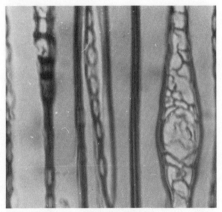

Fig. 3 *Pinus* (t): close-up of horizontal
resin canal

Plate 44
Angiosperm (cross sections)

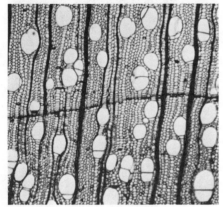

Fig. 1 *Betula* (birch)

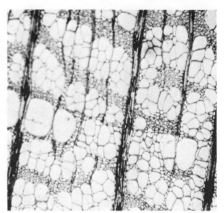

Fig. 2 *Ulmus* (elm)

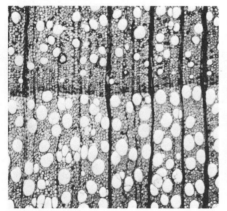

Fig. 3 *Fagus* (beech)

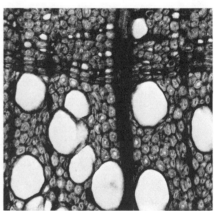

Fig. 4 *Fagus:* close-up

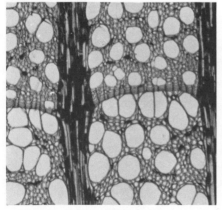

Fig. 5 *Platanus* (sycamore)

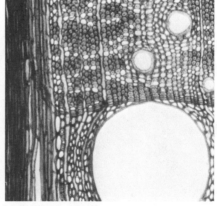

Fig. 6 *Quercus* (oak)

Plate 45
Angiosperms (cross sections) (cont'd.)

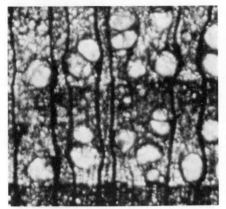

Fig. 1 *Schinoxylon*

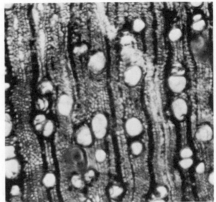

Fig. 2 *Schinoxylon*

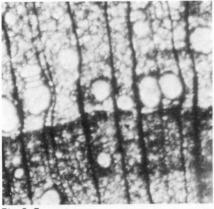

Fig. 3 *Fagara*

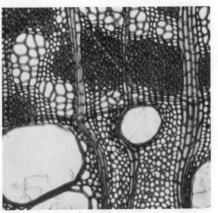

Fig. 4 *Robinia* (locust)

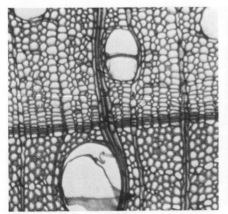

Fig. 5 *Juglans* (walnut)

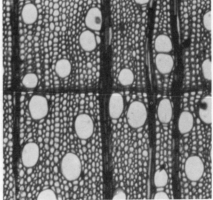

Fig. 6 *Acer* (maple)

Plate 46
Angiosperm (tangential sections)

Fig. 1 *Fagus* (beech)

Fig. 2 *Fagus:* close-up

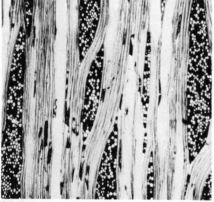

Fig. 3 *Platanus* (sycamore)

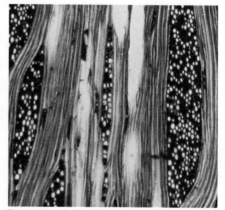

Fig. 4 *Platanus:* close-up

Fig. 5 *Platanus:* close-up of rays

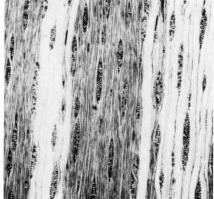

Fig. 6 *Ulmus* (elm)

Plate 47
Angiosperms (tangential sections) (cont'd.)

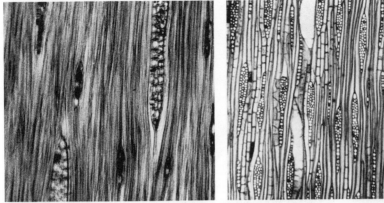

Fig. 1 *Ulmus:* close-up

Fig. 2 *Juglans* (walnut)

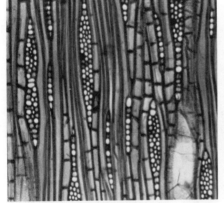

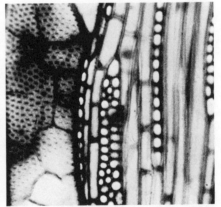

Fig. 3 *Juglans:* close-up

Fig. 4 *Juglans:* close-up

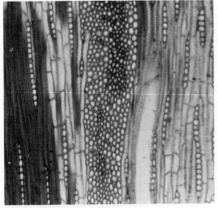

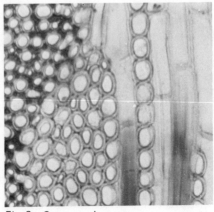

Fig. 5 *Quercus* (oak)

Fig. 6 *Quercus:* close-up

Plate 48
Angiosperms (tangential sections) (cont'd.)

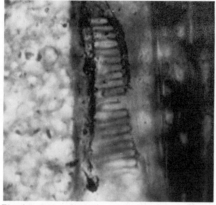

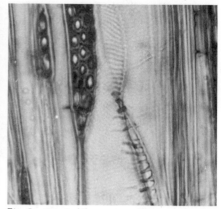

Fig. 1 *Betula* (birch): close-up of
scalariform perforation plate

Fig. 2 *Betula* (?)

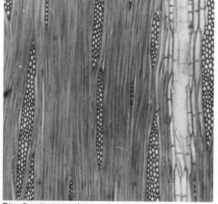

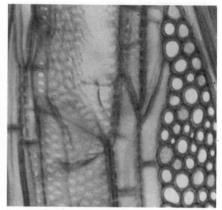

Fig. 3 *Robinia* (locust)

Fig. 4 *Robinia:* close-up

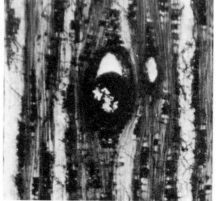

Fig. 5 *Acer* (maple): illustrating spiral and
pitted walls of vessels

Fig. 6 *Schinoxylon:* close-up of gum duct

Petrifaction

Introduction

The identification of fossil wood (petrifaction) is similar to the identification of modern woods, except that fossil woods lack characteristic color, odor, and some other features. The only sure way to identify these woods is to study them in thin section. Thin sections can be easily made with the rock saws and grinding wheels that are usually available to the collector. To study the fine structure of fossil wood for identification requires that the wood be viewed in three planes: cross sectional (transverse), radial, and tangential.

In preparing thin sections, the specimen is first cut or slabbed across the stem, forming a cross or transverse section (Fig. 213a). A second cut, perpendicular to the first and parallel to the ray, constitutes a radial section (Fig. 213c). A third cut, perpendicular to the first two, produces a tangential section (Fig. 213b). Three small blocks trimmed from these cuts (Fig. 214) are then smoothed on a lap wheel (Fig. 215) by grinding in 1200 grit on the surface to be mounted on the glass slide. The smoothed side should not have any small holes or depressions in it; otherwise, air bubbles will form, and the specimen and glass will not adhere correctly. The small block can be mounted to the glass slide with a thermal cement, such as Lakeside 70, or a transparent epoxy resin (Fig. 216). These glass-mounted slides are then ground down on the lap wheels until they are paper thin. This can best be accomplished by initial coarse rapid grinding

117

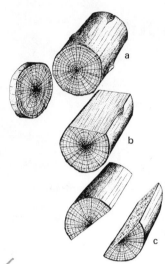

Figure 216. Transverse section being placed on glass slide on heating unit

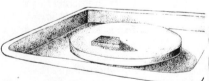

Figure 213. Sections cut in log. a. Transverse (cross), b. Tangential, c. Radial

Figure 214. Trim saw cutting tangential section

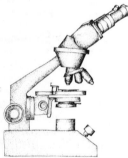

Figure 217. Cover slip being placed on transverse section

Figure 215. Transverse section being smoothed on lap wheel

Figure 218. Thin section slide being studied under microscope

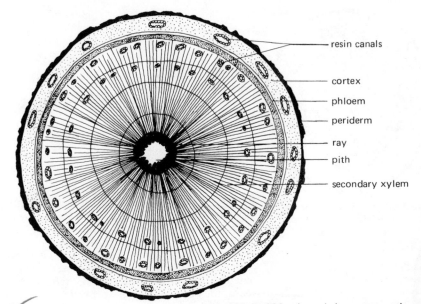

resin canals

cortex

phloem

periderm

ray

pith

secondary xylem

Figure 219. Cross section of
Pinus (pine) stem

with 400 to 600 grit and then progressing to
1200 for final grinding. Be careful not to grind
the slides completely away. A cover slip may be
added for better viewing (Fig. 217). A proper
study of the cellular structure or anatomy of the
wood in the slide specimens requires a microscope or powerful magnifying glass and some
knowledge of the internal structure of plant
stems (anatomy) (Fig. 218).

Stem Anatomy

The stems of plants demonstrate two distinct
types of structure: herbaceous and woody. Understanding the anatomy of the various plant
groups is essential in relating a stem structure to
a particular group. The woody stem is the type
most commonly fossilized. Therefore, the plant
groups under consideration become limited.
Rhizomes of ferns *(Osmundacaulis)* and the
aerial trunks of tree ferns *(Tempskya)* are often
fossilized (see Petrified Ferns). However, the
more common petrified wood is related to the
seed-bearing plants, gymnosperms, and angiosperms. In stem structure, woody dicots and
conifers are similar, whereas monocots have their
own distinct stem anatomy.
In a cross section of a twig of a woody dicot
or conifer, four general zones can be observed
(Fig. 219) beginning with the inner portion of
the stem and moving outward. In the center is

119

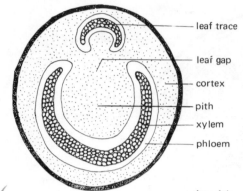

leaf trace

leaf gap

cortex

pith

xylem

phloem

Figure 220. Cross section of fern, illustrating leaf trace and leaf gap

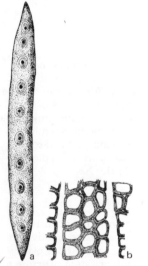

Figure 221. Tracheids. a. Longitudinal (radial) view of tracheid, illustrating shape and pitting, b. Transverse (cross) section of several tracheids

the pith, composed of blockish, thin-walled cells called parenchyma. The pith is surrounded by a broad zone of xylem (wood), the water-conducting tissue of the plant. Outside of the xylem lies the phloem (food-conductive tissue) and the cortex. This latter zone of spongy tissue is composed of parenchyma cells. In older twigs and stems, the protective outermost layer, the epidermis, is replaced by the so-called bark or, botanically, the periderm.

The vascular cylinder or stele, composed of the xylem and phloem, is the conductive region as well as the zone that gives rigidity to the stem.

In fossilization, the material most readily preserved is that most resistant to decay. In plants, this material is composed of cells with secondary thickenings (secondary walls), such as are found in xylem.

Xylem consists of three basic elements: tracheids (Fig. 221), fiber-tracheids, and vessels (Fig. 222). Tracheids are small in cross section, with walls that appear thick. Generally, they are fairly uniform in size. The wood which they compose is called nonporous. Tracheids are common in conifers, other gymnosperms, and ferns.

120

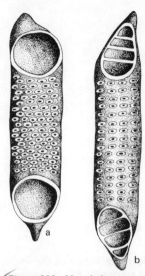

The xylem of angiosperms consists of vessels, plus fiber tracheids in many genera (Fig. 223). Tracheids and fiber tracheids are elongated, tapering cells, whereas vessel shapes vary. Basically, vessels are open to some degree at each end of the cell. These openings are called perforation plates. Both vessels and tracheids have on their walls "dots" known as pits. Water is transferred from one tracheid or vessel to the next through these openings or pits.

Figure 222. Vessel elements. a. Vessel element with alternate pitting and simple perforation plate, b. Vessel element with opposite pitting and scalariform perforation plate

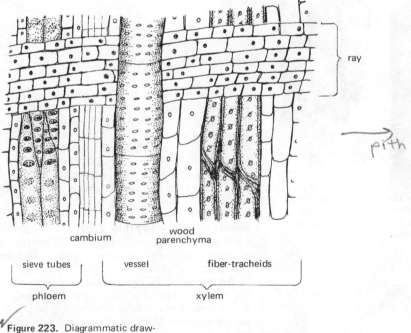

Figure 223. Diagrammatic drawing of radial section of flowering plant stem

121

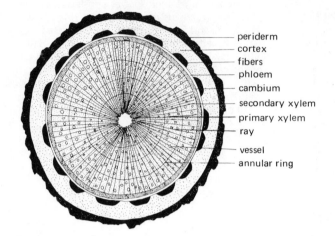

periderm
cortex
fibers
phloem
cambium
secondary xylem
primary xylem
ray

vessel
annular ring

Figure 224. Cross section of angiosperm stem (porous wood)

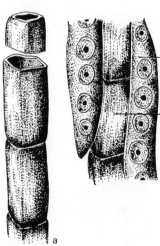

tracheid

parenchyma

Figure 225. Parenchyma cells. a. Transverse (cross) section of a cell, b. Parenchyma cells between tracheids

Because the vessels in the xylem of the angiosperms are larger than tracheids, often they are large enough to be visible to the unaided eye. Thus, angiosperm wood is referred to as porous (Fig. 224).

Thin-walled, generally six-sided parenchyma cells are also found in the xylem (Fig. 225a). These cells are of two types: those that occur throughout the wood (Fig. 225b) and those organized into radially or horizontally aligned layers of cells called rays (Fig. 226). The long axis of the nonray or wood parenchyma is vertical; that of the ray parenchyma is horizontal or radial.

Basically, wood rays are classified as homogenous (all cells alike) (Fig. 226) or heterogenous (cells unlike) (Fig. 227). They are also referred to as uniseriate (single row of cells) and multiseriate (many rows of cells).

An important feature of rays in the gymnosperms is the type of pitting in the lateral walls of the individual ray cells. The number of pits per ray crossing (the area of the ray cell in contact with one tracheid) (Fig. 228), the shape of the pits, and whether they are bordered are features important in classification.

122

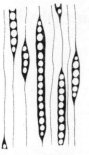

Figure 226. Tangential section of homogeneous uniseriate rays

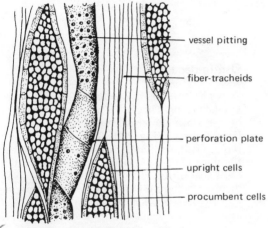

vessel pitting

fiber-tracheids

perforation plate

upright cells

procumbent cells

Figure 227. Tangential section of heterogeneous multiseriate rays with vessels between

ends of tracheids

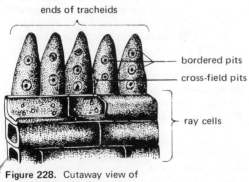

bordered pits

cross-field pits

ray cells

Figure 228. Cutaway view of radial section of ray, showing pitting in crossover fields

Petrified Filicophyta (Ferns)

Figure 229. Living *Osmunda*

Osmundacaulis

Stems or rhizomes of osmundaceous ferns are unique organs. They grow principally upright and bear an open cluster of closely spaced fronds (leaves) on top (Fig. 229). The actual stem is rather small, with its deceptive size arising from the thick mantle of persistent leaf bases and roots that surrounds it. This results in a cone-shaped structure (Pl. 37, figs. 4 and 5) tapering toward its base, which, when fossilized, is assigned to *Osmunda* if Eocene or younger and to *Osmundacaulis* when older.

In many species of living *Osmunda,* the stem extends only a few inches above ground and the frond cluster appears to arise from a low dome of dead leaf material. Two fossil species *(O. carnieri* and *O. braziliensis)* may have grown as small trees.

In cross section the stem appears as a circle of horseshoe-shaped xylem strands surrounding a pith (Fig. 230; Pl. 37, fig. 5). The xylem is enclosed on the outside by the phloem and endodermis. The cortex contains numerous C-shaped petiole (leaf) traces enclosed in boat-shaped wings. These wings may or may not be preserved.

Osmundaceous species of Triassic through Tertiary ages have been reported from various parts of the world. In western North America, species of this genus have been described from the Jurassic Morrison Formation of Utah, the Lower Cretaceous of Queen Charlotte Island, (British Columbia), the Paleocene Fort Union

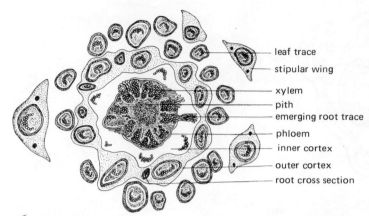

— leaf trace
— stipular wing
— xylem
— pith
— emerging root trace
— phloem
— inner cortex
— outer cortex
— root cross section

Figure 230. Cross section of
Osmundacaulis stem

Figure 231. False trunk of
Tempskya

Formation of North Dakota, and the Eocene
Clarno Formation of Oregon.

Tempskya

Tempskya remains have been reported only from
strata of Lower Cretaceous age. These consist of
club-shaped, corrugated, trunklike structures
several inches in diameter and up to nine feet in
length (Fig. 231). They resemble half-decayed
palm axes. In cross section they are composed of
small, scattered, irregularly circular bodies

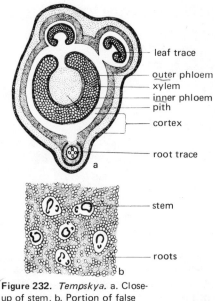

leaf trace

outer phloem
xylem
inner phloem
pith

cortex

root trace

a

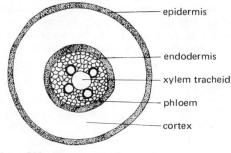

stem

roots

b

Figure 232. *Tempskya.* a. Close-up of stem, b. Portion of false trunk

epidermis

endodermis

xylem tracheid

phloem

cortex

Figure 233. *Tempskya* root

Figure 234. *Tempskya* recon-structions. a. *T. rossica,* b. *T. wesselii*

(roots) enclosing round or lobed figures representing stems (Fig. 232; Pl. 37, fig. 1). The stems are actually a many-branched single stem in which each branch (all resembling the original stem) is encased in a mat of roots (Pl. 37, fig. 2). Stems and roots grew parallel and formed a large, ropelike mass called a false trunk. In cross section, the xylem elements of the roots characteristically form a cross (Fig. 233), whereas *Tempskya* stems consist of the xylem or wood surrounded inside and out by phloem. A three-layered cortex is also present.

The growth habit (Fig. 234) and affinities of this unique fern are not yet completely understood.

126

Petrified Cycadophyta (Cycads)

In many ways the two orders comprising the Cycadophyta, namely, Cycadales (living, sometimes fossil, cycads) (Pl. 38, figs. 1 and 2) and Cycadeoidales (fossil cycads), are similar, particularly with respect to general habit and leaf organization. Trunks of living cycads range from short and squat to tall and columnar (Fig. 235) as do those in Cycadeoidales. The trunks of each are covered protectively with leaf bases that are left after the leaves drop off a short distance from the stem. The basic difference between living cycads and Cycadeoidales are (1) that cones occur at the apex of the stem of the living cycads (Fig. 235) (Pl. 38, fig. 2) but are embedded among the leaf bases in Cycadeoidales (Fig. 236; Pl. 39, fig. 2); and (2) that in Cycadeoidales leaf traces arise directly from the vascular cylinder to the leaves, whereas in living cycads, the traces arise from the vascular cylinder at a level lower than that of the leaf and spiral around nearly the entire diameter of the stem before entering the leaf.

Figure 235. *Stangeria,* a living cycad

Cycadeoidales

Fossil remains of members of this order are found in many parts of the world, particularly in South Dakota, Wyoming, Idaho, Utah, and California. They had their origin during the Triassic or before, persisted throughout the Mesozoic, and became extinct toward the end of the Cretaceous. Most specimens of the genus *Cycadeoidea* display a squat trunk (Pl. 28, fig. 2)

— leaf bases
— cones

Figure 236. Reconstruction of *Cycadeoidea*

127

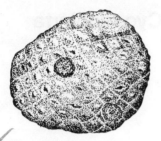

Figure 237. Stem of *Cycadeoidea*

with a dense covering of leaf bases (Fig. 237). Attached mature leaves have not been found, although immature leaves have (Fig. 238). Those that have been discovered indicate that the leaves were arranged in a crown at the top of the stem.

Specimens of another genus of this order known as *Monanthesia* have been collected from the San Juan Basin, New Mexico, and near Moab, Utah. They consist of a columnar stem having a uniform diameter of up to 20 inches (Fig. 239; Pl. 38, fig. 5). These trunks have spiral, resistant, triangular leaf bases with a cone in the axil of each leaf.

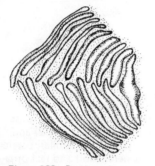

Figure 238. Portion of an immature leaf of *Cycadeoidea*

Figure 239. Stem of *Monanthesia*

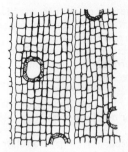

Petrified Coniferophyta (Conifers)

Figure 240. Normal resin canals

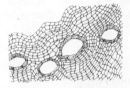

Figure 241. Traumatic resin canals

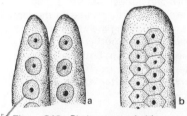

Figure 242. Pitting on tracheids. a. Uniseriate with rounded pits, b. Biseriate with angular, alternate pits

Figure 243. Crassulae around pits

Because of the similarity of petrified conifer woods, identification is often very difficult. Answers to the following questions help to distinguish them:

1. Are annual rings present or absent? If present, how distinct are they?
2. Are resin canals present or absent? If present, are they a product of normal growth (Fig. 240), as in pine, or are they traumatic, products of injuries (Fig. 241), as in *Sequoia?*
3. Is wood parenchyma present or absent? If present, how much?
4. Are the tracheids pitted? Are the pits round or angular (Fig. 242)? Are the pits in two or more rows on the tracheids? If in more than one row, are they opposite or alternate? Are they touching? Do thickenings (known as crassulae or bars of Sanio) occur between the pits (Fig. 243)?

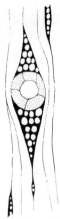

Figure 244. Tangential view of horizontal resin canal in fusiform ray

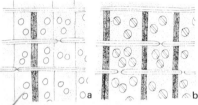

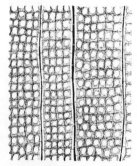

Figure 245. Crossover fields. a. Oopores, b. Oculipores

5. Are rays uniseriate or multiseriate? Do the multiseriate rays in conifers contain a horizontal resin canal (Fig. 244)? What is the height of the rays? (The height is determined by the number of vertical cells composing the ray.)
6. How many pits are in the crossover field? Do they have borders (oculipores) (Fig. 245b) or not (oopores) (Fig. 245a)?

The following discussion of genera related to petrified conifer wood is not comprehensive. The genera included are those identified from the western United States. Formations and ages in which they are most commonly encountered in this region are given.

Araucariaceae

Araucarioxylon (Dadoxylon)

Specimens (Pl. 34, fig. 1) assignable to this genus are common in formations of Triassic (Chinle and Shinarump) and Jurassic (Morrison) ages. These fossils are supposedly related to the family Araucariaceae, whose living members are presently growing in the Southern Hemisphere. Resin canals are absent in both normal and wounded xylem in this genus (Fig. 246; Pl. 34, fig. 2). Bordered pits are alternate and angular

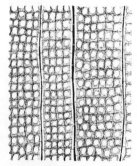

Figure 246. Cross section of *Araucarioxylon*

130

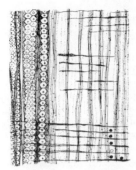

Figure 247. Radial section of *Araucarioxylon*

(polygonal) when they occur in two or more rows (Pl. 34, fig. 3); if uniseriate, they are flattened above and below (Fig. 247). Bars of Sanio and wood parenchyma are rare or absent in this genus. Rays are generally uniseriate (Fig. 248).

Woodworthia

Specimens of *Woodworthia* are common in the Triassic Chinle Formation. On the surface of these specimens are many small scars and a few larger ones (Fig. 249; Pl. 39, fig. 6). The small scars possibly represent the short shoots; the larger, the leaf or branch traces.

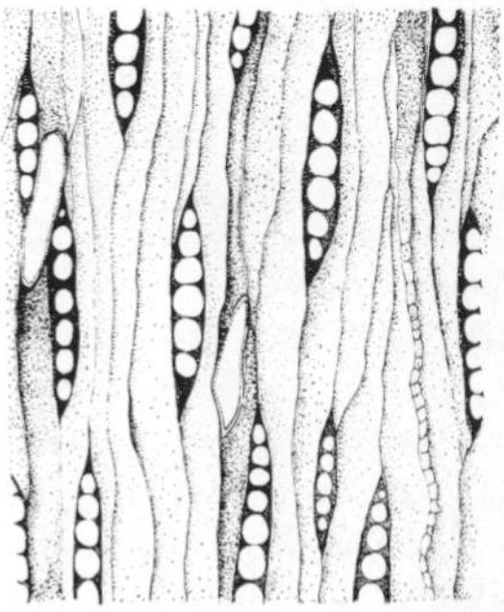

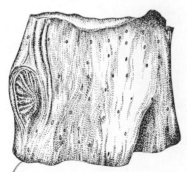

Figure 248. Tangential section of *Araucarioxylon*

Figure 249. *Woodworthia*

131

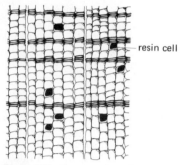

Figure 250. Cross section of
Sequoia (redwood)

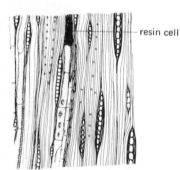

Figure 251. Tangential section of
Sequoia (redwood)

Taxodiaceae

Sequoia (redwood)

Petrified wood assignable to this genus has been
reported from Tertiary floras from several localities such as the Ginkgo State Park area in Washington and Yellowstone National Park. Few
anatomical characters serve to separate *Sequoia*
from *Taxodium, Sequoiadendron,* and *Metasequoia;* therefore, no separation will be at-
tempted in this book.

Growth rings in *Sequoia* wood are usually
narrow, although this may vary (Fig. 250). The
transition from late (summer) wood to early
(spring) wood is generally abrupt. Resin ducts,
when present, are of the traumatic type only.
Rays are generally uniseriate, sometimes biseriate. Wood parenchyma contains resin deposits
(Fig. 251) known as resin cells (Pl. 40, figs. 5 and
6). Bordered pits occur in one or two rows on
the tracheid. Sometimes as many as four may be
observed. Crassulae (bars of Sanio) are infre-
quent.

The genus *Cupressinoxylon* has been used for
fossil woods similar to *Sequoia* but lacking the
traumatic resin ducts, whereas those displaying
traumatic ducts are often placed in the form
genus *Sequoioxylon.*

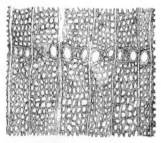

Figure 252. Cross section of
Pityoxylon

132

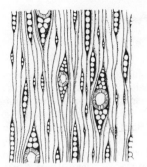

Figure 253. Tangential section of *Pityoxylon*

Pinaceae

Pityoxylon

Specimens of *Pityoxylon* contain normal resin canals both vertically and horizontally (Figs. 252-254; Pl. 34, figs. 4-6). This Mesozoic genus exhibits anatomical features associated with those of pines, spruces, larch, douglas fir, and others.

Pinus (pine)

Growth rings in pine wood are usually distinct and often wide. Both horizontal and longitudinal resin ducts are present (Figs. 255 and 256). Rays are of two types: a normal, generally uniseriate ray that is sometimes biseriate and a fusiform type that includes a horizontal resin duct.

Bordered pits occur generally as a single row, rarely as two, on the tracheid. Crassulae (bars of Sanio) are well developed.

Fossil pine wood has been reported from the Tertiary of the western United States (Pl. 41, Pl 43).

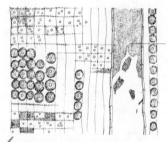

— resin canal

Figure 254. Radial section of *Pityoxylon*. Note large vertical resin canal.

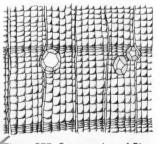

Figure 255. Cross section of *Pinus* (pine)

133

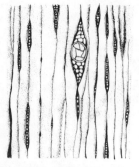

Figure 256. Tangential section of *Pinus* (pine)

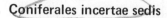

Coniferales incertae sedis

Xenoxylon

This genus is characterized by having very large pits on the radial walls of the tracheids. These pits are close together and are generally flattened above and below (Pl. 35, fig. 5). Resin canals are lacking (Fig. 257). Rays are uniseriate, and generally only one large simple pit occurs in the cross field (Pl. 35, figs. 2 and 4). *Xenoxylon morrisonense,* described from the Morrison Formation of Utah, has conspicuous septa in its tracheids (Fig. 258).

Cupressinoxylon

This genus has been reported throughout various parts of the world from rocks Jurassic to Recent in age. However, most specimens assigned to this genus from the western United States are from the Cretaceous.

This generic name is applied to fossil woods exhibiting the following structure:

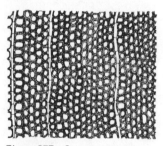

Figure 257. Cross section of *Xenoxylon morrisonense*

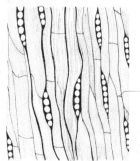

—— septate tracheid

Figure 258. Tangential section of *Xenoxylon morrisonense*

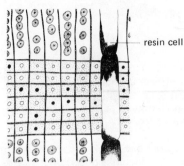

resin cell

Figure 259. Radial section of *Cupressinoxylon*. Note resin cells.

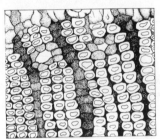

Figure 260. Cross section of *Protopiceoxylon*

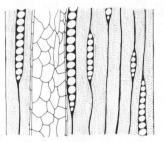

Figure 261. Tangential section of *Protopiceoxylon*

1. narrow, well-defined annual rings
2. abundant wood parenchyma that often contains resin (These cells can be recognized by their dark contents [resin] even in transverse section.) (Fig. 259; Pl. 42, fig. 1)
3. resin canals lacking, except in areas of wounding
4. bordered pits usually separate and circular; if in more than one row on tracheid, then opposite

Cupressinoxylon resembles *Cedroxylon* except in containing more wood parenchyma, particularly in the spring wood.

Protopiceoxylon

This genus is characterized by conifer wood that has normal vertical resin canals (Fig. 260; Pl. 42, figs. 2-4) but lacks the horizontal resin canals in the normal wood (Fig. 261). Pits on the tracheids are generally separate and circular. Well-developed bars of Sanio are present on the tracheids.

This genus has been reported from the Morrison Formation of Jurassic age.

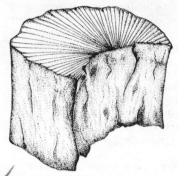

Figure 262. *Schilderia*

Gnetales (?)

Schilderia

Schilderia is easily recognized in the field by its numerous, conspicuous multiseriate rays that are clearly evident in cross section (Fig. 262). The prominent rays exhibit a pattern like the rays of the sun, radiating from the pith region of the stem (Pl. 29, fig. 6).

The relationship of *Schilderia* to other members of the plant kingdom is uncertain; its association with the Gnetales is therefore only suggested. It has been collected from the Triassic Chinle Formation.

136

Petrified Anthophyta (Angiosperms)

Dicotyledonae

Definite determination of fossil woods assignable to the Dicotyledonae (dicots) is often difficult—under even the most favorable conditions—for several reasons. First is imperfect preservation. If the diagnostic characters are not preserved, then exact identification is unlikely. Furthermore, the number of living dicot genera is high, often making it impossible to relate the fossil to a modern genus, particularly when the fossil genus was more important in the fossil flora than is its modern correlative in the existing flora. Thus, the relationship of a form genus of fossil wood is generally not narrowed beyond the living family. Again the variation of wood structure within a family may be so wide that more exact placement of the fossil wood is not always possible. Fourth, the petrified wood often may represent an extinct form without modern relatives.

But despite these difficulties, it is usually possible to place fossil dicot woods within a broad group by answering the following questions:

1. Are annual rings present or absent? If present, are they distinct or indistinct?
2. What is the arrangement of vessels as viewed in cross section? That is, are they scattered (diffuse porous) (Fig. 264), or are they generally aligned along the annual ring (ring porous) (Fig. 265)?

Figure 263. Living *Acer* (maple)

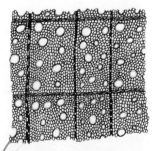

Figure 264. Diffuse porous

137

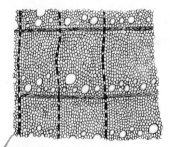

Figure 265. Ring porous

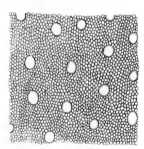

Figure 266. Solitary vessels

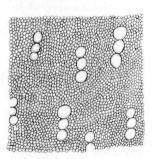

Figure 267. Multiple vessel chains

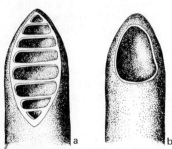

Figure 268. Perforation plates. a. Scalariform, b. Simple

3. Is each vessel isolated (solitary) (Fig. 266), or are the vessels connected (multiples) (Fig. 267)?
4. Is the distribution of the vessels even or uneven?
5. What is the number of vessels per square millimeter?
6. What is the average diameter and the average length of the vessel elements?
7. What type of perforation plate or opening is on the ends of the vessels? That is, are the openings scalariform (ladderlike) (Fig. 268a) or simple (having only one opening) (Fig. 268b), or both? If scalariform, how many bars are on the plate?
8. Are tyloses present or absent in the vessel element (Fig. 269)? (Tyloses are outgrowths or protrusions from parenchyma cells surrounding the vessel. These enter into the vessel when it becomes inactive or injured.)
9. Is wood or vertical parenchyma present or absent? If present, how much is there and how is it distributed?
10. Are the rays heterogenous or homogenous? Are they multiseriate, uniseriate, or both?

The following genera represent some of the more commonly collected fossil dicot woods in the western United States:

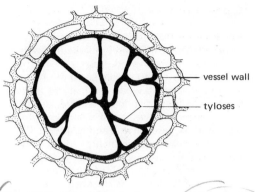

vessel wall

tyloses

Figure 269. Cross (transverse) section of vessel filled with tyloses

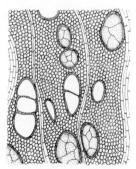

Figure 270. Cross section of *Paraphyllanthoxylon*

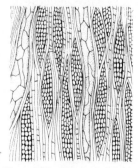

Figure 271. Tangential section of *Paraphyllanthoxylon*

Lauraceae (?)

Paraphyllanthoxylon. Growth rings in the wood of this genus appear to be lacking. The wood is diffuse porous, composed of solitary vessels, or multiples of two to five (Fig. 270; Pl. 36, figs. 1 and 2) with simple perforation plates. Tyloses are abundant in the numerous vessels. Xylem or wood parenchyma is sparse. Rays are generally multiseriate, a few uniseriate types being present (Pl. 36, figs. 3 and 4). Multiseriate rays are two to four cells wide and are heterogenous (Fig. 271). Septate fiber-tracheids are common throughout the wood.

Paraphyllanthoxylon arizonense is known from the Upper Cretaceous of Arizona, *P. idahoense* has been reported from the Lower Cretaceous Wayan Formation of Idaho, and *P. utahense* has been described from the Lower Cretaceous Cedar Mountain Formation of Utah.

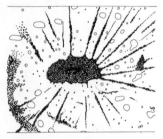

Figure 272. Cross section of
Forchhammerioxylon

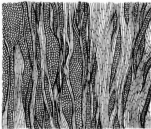

Figure 273. Tangential section of
Forchhammerioxylon

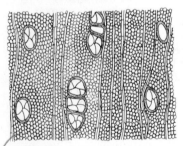

Figure 274. Cross section of
Heveoxylon

Figure 275. Tangential section of
Heveoxylon. Note tyloses in
vessel.

Forchhammerioxylon. The form genus *Forch-hammerioxylon* has closely marked growth rings and diffuse porous wood. Vessels are relatively small and short with simple perforation plates. Tyloses are absent; wood parenchyma is present. Among the most distinctive features of this wood are the successive wedges of xylem and phloem separated by parenchyma and large rays (Fig. 272). The rays are prominent, wide, and high. Fibers in the wood are septate (Fig. 273).

The close relative of this form genus, the living *Forchhammeria,* is confined to the Americas north of the Equator (from California to Guatemala and the West Indies). *Forchhammerioxylon* has been reported from the Eocene of Eden Valley, Wyoming.

Euphorbiaceae

Heveoxylon. Growth rings are indistinct or absent in this fossil wood reported from Eden Valley, Wyoming. The wood is ring porous consisting of solitary vessels or, sometimes multiples of two to nine (Fig. 274). Perforation plates on the vessels are simple, and vessel-wall pitting is alternate. Thick-walled tyloses are abundant. Wood parenchyma is sparse. Rays are both uniseriate and biseriate, and all are strikingly heterogenous (Fig. 275).

Heveoxylon is similar to the woods of the living species *Hevea microphylla,* which is related to *Hevea braziliensis,* the chief rubber tree of commerce. *H. microphylla* grows in northern South America, from the Amazon Basin to Venezuela.

140

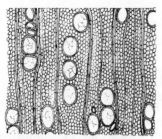

Figure 276. Cross section of *Fagara*

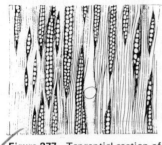

Figure 277. Tangential section of *Fagara*

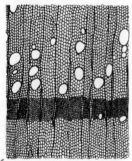

Figure 278. Cross section of *Schinoxylon*

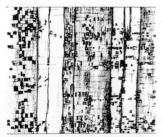

Figure 279. Radial section of *Schinoxylon*

Rutaceae

Fagara. Petrified woods assigned to *Fagara* have been reported from Eden Valley, Wyoming. Growth rings are present but may be indistinct (Pl. 39, fig. 4). Vessels are diffuse porous and occur both as solitary vessels and in radial chains of two to nine (Fig. 276; Pl. 45, fig. 3). They have simple perforation plates and may contain tyloses. Wood parenchyma is paratracheal and, depending upon the species, may be sparse or abundant. Rays are heterogenous and are uniseriate or biseriate, rarely triseriate (Fig. 277).

Species of *Fagara* presently live in the pantropical region. Members of this genus grow in southern Florida.

Anacardiaceae

Schinoxylon. This fossil wood has distinct growth rings and is diffuse porous (Fig. 278; Pl. 45, figs. 1 and 2). Two vessel types are present: One is large, solitary, sometimes in multiples of 2 to 4, and contains tyloses; the second is small and tracheidlike in appearance. They usually occur in multiples of up to 25 vessels. Wood parenchyma is absent (Fig. 279). Fibers are thin-walled and nonseptate. Rays are heterogenous and are generally uniseriate, although they may infrequently be biseriate.

Schinoxylon is characterized by having numerous, horizontal gum ducts that are somewhat elongated vertically (Pl. 48, fig. 6). Rays containing gum ducts are considerably enlarged (Fig. 280).

This form genus is similar to the living *Schinus* (peppertree) of the Anacardaceae, a family well represented as compressions from the Upper Cretaceous to the present throughout the world. Other described wood genera tentatively assigned to the Anacardiaceae are *Anacardioxylon* from the Tertiary of California and *Edenxylon* (which also contains horizontal gum ducts) from Eden Valley, Wyoming.

141

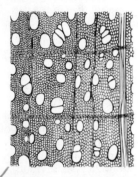

Figure 280. Tangential section of *Schinoxylon*. Note gum duct.

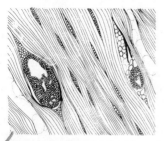

Figure 281. Cross section of *Acer* (maple)

Acer (maple). Fossil woods related to *Acer* are known from the Cretaceous to Recent under various form-generic names. Growth rings of *Acer* woods are fairly distinct and often delimited by two or three rows of thick-walled fibers (Fig. 281; Pl. 45, fig. 6). The vessels are normally solitary, although they sometimes occur in multiples. They are quite uniform in size and are evenly distributed as viewed in cross section. They have simple perforation plates, and tyloses are absent. Wood parenchyma is sparse. Rays are homogenous, nonstoried, and from one to seven cells wide (Fig. 282; Pl. 48, fig. 5). Uniseriate rays are fairly common. Wider rays (five to seven cells wide) appear on the tangential surface as short, crowded lines visible to the unaided eye.

Acer wood may superficially resemble *Betula* (birch) wood, but in *Betula* the perforation plates are scalariform rather than simple as in *Acer* (Fig. 282). Spiral thickenings are also present on some of the vessels of *Acer,* but absent from those of *Betula* (Pl. 48, fig. 5). The ray cells of *Betula* as viewed in tangential section have a flattened appearance (Fig. 283), whereas the ray cells of *Acer* are more rounded (Fig. 282).

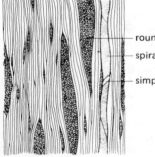

— rounded ray cells

— spiral thickenings

— simple perforation plate

Figure 282. Tangential section of *Acer* (maple)

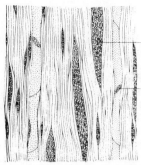

— elongated ray cells

— scalariform perforation plate

Figure 283. Tangential section of
Betula (birch)

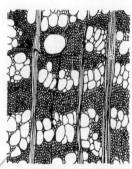

Figure 284. Cross section of
Ulmus (elm)

Ulmaceae

Ulmus (elm). Growth rings are distinct and fairly wide in this wood, the transition between the early and late wood being rather abrupt (Fig. 284). The wood is ring porous (Pl. 44, fig. 2). Vessels are solitary, with simple perforation plates. Tyloses occur in the vessels of some species but not in all. Wood parenchyma is present. Rays are homogenous and are one to seven (usually four to six) cells wide (Fig. 285; Pl. 46, fig. 6 and Pl. 47, fig. 1). Uniseriate rays are rare. *Ulmus* wood is known from Miocene strata near Vantage, Washington; and *Ulmoxylon simrothii,* which demonstrates a close resemblance to modern elms, has been reported from the Pliocene of California.

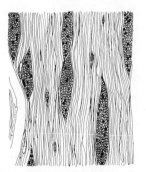

Figure 285. Tangential section of
Ulmus (elm)

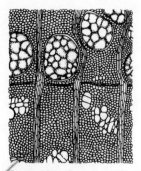

Figure 286. Cross section of *Robinia* (locust)

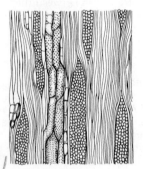

Figure 287. Tangential section of *Robinia* (locust)

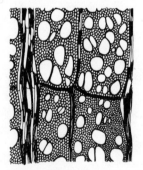

Figure 288. Cross section of *Platanus* (sycamore)

Leguminosae

Robinia (locust). Living *Robinia* are shrubs or medium-sized trees. Fossil woods of *Robinia* have been described from the Miocene of Montana and the Pliocene of California.

The wood is ring porous with distinct growth rings (Pl. 45, fig. 4). Vessels are relatively few and solitary (Fig. 286). Wood parenchyma is abundant. Fibers are nonseptate and thick walled, mostly irregular in cross section. Rays are unstoried and homogenous to somewhat heterogenous (Fig. 287; Pl. 48, figs. 3 and 4).

Platanaceae

Platanus (planetree or sycamore). Growth rings of this wood are distinct, and the wood is diffuse porous (Fig. 288; Pl. 44, fig. 5). Vessels are small, numerous, frequently crowded, distinctly solitary, or in irregular groups of two to five. Perforation plates are simple or scalariform. Tyloses are absent. Pits are not crowded on the vessel walls. Thick-walled fiber-tracheids are nonseptate, and wood parenchyma is relatively sparse. Rays are unstoried, homogenous, wide, and conspicuous (Pl. 46, figs. 3-5). They are only multiseriate (3-14 seriate) (Fig. 289).

Fossil *Platanus* wood has been reported from the Miocene of Nevada City, California; Yellowstone National Park; and Ginkgo State Park, Washington.

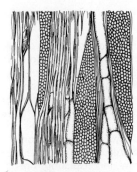

Figure 289. Tangential section of *Platanus* (sycamore)

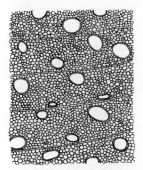

Figure 290. Cross section of *Myrica*

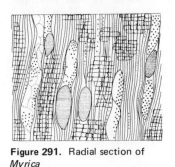

Figure 291. Radial section of *Myrica*

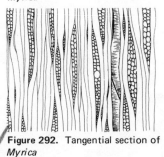

Figure 292. Tangential section of *Myrica*

Myricaceae

Myrica. Species of this genus are currently growing throughout temperate and tropical regions of both the Eastern and Western hemispheres. Compressions of leaves, flowers, and seeds of *Myrica* have been reported from floras of Cretaceous and Tertiary ages. Fossil woods assigned to *Myrica* have been collected from the Eocene Green River Formation in Eden Valley, Wyoming.

The fossil wood of *Myrica* is diffuse porous and has indistinct growth rings (Fig. 290). Vessels are usually solitary, although an occasional multiple of two vessels may be observed. Scalariform perforation plates are present on the vessels (Fig. 291), some of which contain large tyloses. Wood or vertical parenchyma is abundant. Heterogenous rays are bi- or triseriate, mostly biseriate (Fig. 292). These rays are composed of square or upright cells interspersed with procumbent cells.

145

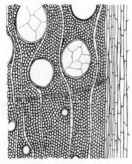

Figure 293. Cross section of *Quercus* (oak)

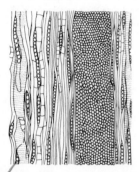

Figure 294. Tangential section of *Quercus* (oak)

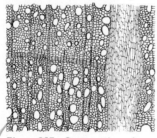

Figure 295. Cross section of *Fagus* (beech)

Quercus (oak). Fossil woods and leaves of oak have been reported throughout the world from Cretaceous age to the present. Wood has been described from the Miocene of Yellowstone, Idaho, and California. Typically it is ring porous (Fig. 293; Pl. 44, fig. 6) with two types of rays (Pl. 47, figs. 5 and 6). The first type, consisting of large, broad rays, is usually accompanied by the second type, which consists of numerous small, narrow, inconspicuous uniseriate rays (Fig. 294). Tyloses are present in the vessels.

Fagus (beech). Wood of *Fagus* has distinct growth rings and is diffuse porous, containing a relatively large number of scattered vessels (Fig. 295; Pl. 44, figs. 3 and 4). Tyloses are sometimes present in the center or heart wood. Rays are of two types (Pl. 46, figs. 1 and 2): broad, 15 to 25-plus cells wide; and narrow, 1 to 5 seriate. Broad rays, plainly visible to the unaided eye, are separated by narrow rays (Fig. 296). They appear on the tangential surface as short, somewhat widely spaced staggered lines that are visible without magnification. Wood parenchyma is abundant, and pitting is crowded on the vessels.

Fagus wood is similar to *Quercus* (oak), particularly in the two sizes of rays. They differ in that beech wood is diffuse porous and oak is ring porous.

Fagus wood has been reported from the Miocene of Yellowstone National Park.

146

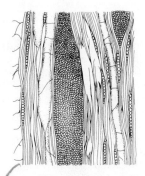

Figure 296. Tangential section of *Fagus* (beech)

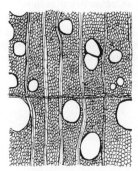

Figure 297. Cross section of *Juglans* (walnut)

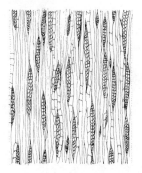

Figure 298. Tangential section of *Juglans* (walnut)

Juglandaceae

Juglans (walnut). Petrified woods assignable to the family Juglandaceae have been reported from several localities in the Northern Hemisphere in both the Old and New worlds. The genera *Juglans, Carya* (hickory), and *Pterocarya* (wingnut) have been described from the Oligo-Miocene strata of Oregon and the Miocene near Vantage, Washington.

Woods of *Juglans* have distinct growth rings and are diffuse to semi-ring porous (Fig. 298; Pl. 45, fig. 5). The vessels are chiefly solitary. They are evenly distributed and closely placed, although there is an abrupt change in vessel size between the outer late wood and the early wood of the succeeding ring. Those of the early wood are visible to the unaided eye and gradually decrease in size toward the outer margin of the ring. The wood may appear semi-ring porous. Wood parenchyma is present and may sometimes be enlarged and crystalliferous. Rays are both homogenous and heterogenous (Fig. 298). They are nonstoried and are usually one to five cells wide, the uniseriate rays being rather limited in number (Pl. 47, figs. 2-4).

Monocotyledonae

In a monocotyledonous stem, the individual bundles containing the vascular portion are scattered instead of being arranged in a cylinder as in herbaceous or woody dicot stems. The ground tissue in which the bundles are enclosed is composed of thin-walled parenchyma cells. These cells are utilized by the plant for temporary food accumulation. It is from these cells in sorghum and sugar cane that sugar is obtained.

Palmoxylon

Palmoxylon is the genus designated for the petrified trunks of palms and other arborescent (treelike) monocots (Fig. 299). They have been collected from many of the western states. Localities have been reported in Texas, Arizona, Utah, and California, and from the famous Eden Valley of Wyoming.

Palmoxylon axes are composed of scattered oval or bottle-shaped vascular bundles (Fig. 300; Pl. 33, fig. 2). Each bundle consists of a small vascular portion of one to four (usually two) large vessels and a darker portion that makes up the fibrous bundle cap (Pl. 33, fig. 3). Depending upon the species, smaller, roundish bundles composed only of fibers (called fibrous bundles) may be observed.

Petioles and roots are often found attached to the trunks. The petioles are C-shaped and tapered, broadening downward.

Figure 299. *Palmoxylon*

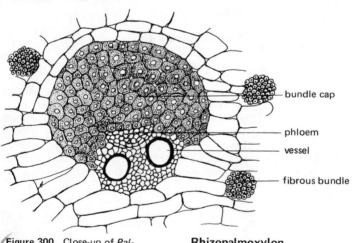

— bundle cap

— phloem

— vessel

— fibrous bundle

Figure 300. Close-up of *Palmoxylon* bundle

Rhizopalmoxylon

Rhizopalmoxylon (Pl. 38, fig. 3) is the form genus for petrified roots referable to the family Palmae. These roots are small and their stele is composed of several individual vessels in from one to three concentric rings near the root center (Fig. 301). External to these is an endodermis, cortex, and, when preserved, epidermis. Phloem occurs between the outermost ring of vessel elements. Phloem, however, is rarely preserved.

148

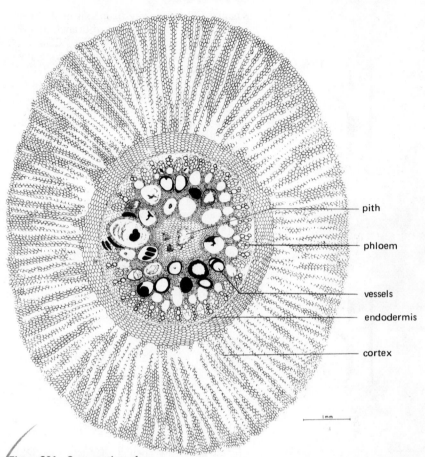

pith

phloem

vessels

endodermis

cortex

1 mm

Figure 301. Cross section of
Rhizopalmoxylon blackii Tidwell.
Note the thick-walled conjunctive
tissue. (Cells slightly enlarged)
(After Tidwell, Medlyn, and
Thayn, 1972)

149

Glossary

Abaxial—directed away from axis or shoot of plant (e.g., lower surface of branch in relation to main stem) (*see* Fig. 302a)

Acuminate apex—apex tapering to point, the sides more or less concave (*see* Fig. 328)

Acute apex—rather sharp pointed apex, not drawn out (*see* Fig. 328)

Adaxial—directed toward axis or shoot (e.g., upper surface of branch in relation to main stem) (*see* Fig. 302b)

Aerate—to allow air to circulate through

Aerenchyma—large-celled tissue containing air spaces for gaseous exchanges of oxygen and carbon dioxide (*see* p. 51)

Affinity—affiliation; relation between plants based on similarity of structure and origin

Aggregate—cluster (i.e., cluster of cells)

Alga (plur.: algae)—chlorophyll-bearing thallo-phytes that usually grow in water and are known as seaweeds or waterweeds (*see* p. 11)

Ament—catkin or catkinlike

Anastamosing—said of veins fusing and dividing to form a netlike pattern (*see* Fig. 79)

Angiosperm—flowering plant or, more specifically, plant with seeds borne in ovary (*see* p. 19)

Apex (plur.: apices)—tip of pinnule or leaf (*see* Figs. 82, 83)

Apical—pertaining to apex

Apophysis—exposed surface of an organism (*see* p. 90)

Appress—press to, or against, surface

Arborescent—treelike

Aristate apex—awl-shaped apex (*see* Fig. 328)

Asymmetrical—without symmetry

Attenuate apex—apex drawn out to slender point, the sides straight (*see* Fig. 328)

Auriculate base—base with lobe resembling ear on each side of petiole or midvein (*see* Fig. 331)

Awl-shaped—slender, sharp pointed

Axil—upper angle formed between stem or branch and its branch, twig, or leaf (*see* Fig. 302c)

Bars of Sanio—markings on radial wall of tracheid in some conifers (*see* p. 129)

Bifurcate—to divide into two branches

Binomial—biological species name consisting of two terms (*see* p. 21)

Biseriate—in two rows (e.g., biseriate rays consist of cells in two rows)

Blade—flat, expanded part of leaf (*see* p. 95)

Brackish water—fresh water mixed with salt; briny

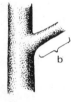

Figure 302. a. Abaxial, b. Adaxial, c. Axil

150

Figure 303. Cirque

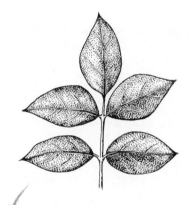

Figure 304. Compound leaf

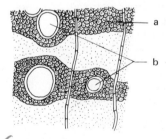

Figure 305. Conjunctive wood parenchyma. a. Parenchyma cells, b. Vessels

Figure 306. Convex surface

Bract—modified leaf or scale
Bulbil—small bulb
Bundle (vascular bundle)—strand of specialized cells that conducts both water and food in plants
Bundle cap—fibrous structure on top of vascular bundle
Bryology—study of bryophytes (*see* p. 13)

Calyx (plur.: calyces)—external, usually green or leafy part of flower consisting of sepals
Camptodromic (camptodromous)—said of secondary or lateral veins curving toward margins without forming loops (cf. anastomosing)
Carbonate—composed of $CaCO_3$ or limestone
Carboniferous—Carboniferous Period (*see* p. 26)
Chlorophyll—green coloring material in plants that produces sugars from carbon dioxide, water, and oxygen in presence of sunlight (i.e., during photosynthesis)
Circinate vernation—coiled arrangement in immature leaves of ferns, commonly called fiddleheads
Cirque—steep, spoon-shaped depressions on sides of mountains; produced by glacial erosion (*see* Fig. 303)
Climax cover—(in plant ecology) final vegetational type in plant succession
Compound leaf—leaf in which blade is divided to midrib, forming two or more leaflets on a common axis (*see* Figs. 304, 327)
Compression—fossil-plant remains flattened on surface of rock layer
Conductive tissue—tissue that conducts water upward (xylem or wood) and food downward (phloem) in stem
Cone scales—stiff scales comprising female conifer cones (seeds are borne on scales) (*see* Fig. 141b)
Conifer—(*see* p. 82)
Conjunctive wood parenchyma—connected or coalesced parenchyma (*see* Fig. 305)
Constrict—to contract or curve inward
Convex—curved upward or outward (*see* Fig. 306)
Cordaitales—group of ancient gymnosperms: tall trees with straplike leaves (*see* p. 83)
Cordate—heart shaped (*see* Fig. 329)
Cortex—tissues external to phloem
Cotyledon—seed leaf of embryo, or earliest leaf arising from seed
Craspedodromic (craspedodromous)—said of veins going from midvein to margins without

151

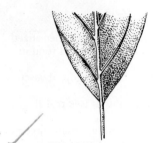

Figure 307. Cuneate base

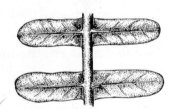

decurrent

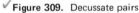

Figure 308. Decurrent pinnule

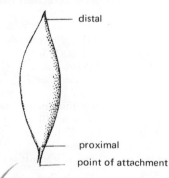

Figure 309. Decussate pairs

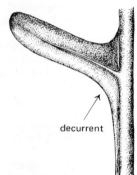

distal

proximal

point of attachment

Figure 310. Distal and proximal ends with point of attachment

dividing (cf. anastomosing)

Crassulae—Bars of Sanio

Crenate—having margin cut into rounded scallops (see Fig. 330)

Cross or transverse section—cut made at right angles to long axis of stem

Cuneate—wedge shaped (see Figs. 307, 331)

Cuticle—outermost layer of plant epidermis; composed of cutin, which repels water and therefore helps retain water within the plant

Cutin—combination of fatty acids and waxes that comprises chief ingredient of cuticle; repels water

Cycadales—order of gymnosperm represented by single surviving family (Cycadaceae) of tropical plants resembling palms but reproducing by cones, not flowers

Deciduous—falling off or shed seasonally

Decurrent—said of leaf, leaflet, or pinnule extending down stem or rachis (see Fig. 308)

Decussate—alternating in pairs at right angles (see Fig. 309)

Dehiscent—split open

Deltoid—delta shaped, triangular (see Fig. 329)

Dentate—having teeth or pointed conical projections (see Fig. 330)

Diaphragm—partition separating tissues

Dichotomy—division into two exclusive parts

Dicot—informal abbreviation for dicotyledonae (see p. 21)

Diffuse porous—vessels scattered throughout wood (see p. 137)

Digitate—like fingers

Distal—a point or area farthest from the point of attachment (see Fig. 310)

Diversity—variety

Doubly serrate—(leaf margin) having smaller serrations imposed on larger ones (see Figs. 311, 330)

Drip point—long, attenuate apex on some plants, particularly in regions of heavy rainfall, allowing for more efficient runoff from the leaf

Elator—club-shaped, water-absorbing appendage or band attached to spores of *Equisetum* (see p. 15)

Elliptic—oval shaped with rounded ends (see Fig. 329)

Emarginate—having margin notched (see Fig. 328)

Encrust—to coat or cover with crust (e.g., a carbonate)

Endodermis—layer of specialized cells, often with

Figure 311. Doubly serrate margin

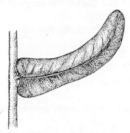

Figure 312. Falcate pinnule

thickened walls, separating pericycle from cortex in roots and, infrequently, in stems

Entire margined—having a smooth or unbroken margin (*see* Fig. 330)

Epiphyte—nonparasitic plant that grows on other plants; air plant

Exsert—to project beyond surrounding structure

Falcate—curved like a sickle (*see* Fig. 312)

Fascicle—cluster or bundle of leaves, flowers, etc. (commonly used in relation to needles of pine)

Fauna—animal life

Fiber—long, slender, thick-walled, tapering plant cell serving to protect and support the plant body (*see* Fig. 313)

Fiddlehead—(see circinate vernation)

Filament—threadlike plant body of some algae

Flora—plant life

Formation—(geology) group of similar rocks or strata; mappable unit

Form genus—name applied to plant parts having similar morphological structure (*see* p. 24)

Frond—leaf (*see* p. 61)

Frond cluster—group of fronds from same rhizome

Fructification—formation of reproductive or fruiting body

Gamete—unisexual body that, upon fusion with another gamete, gives rise to sporophyte plant

Gametophyte—gamete-producing plant

Genus (plur.: genera)—subdivision, in plant and animal classification, in which members have common distinguishing characters (*see* species).

Girdling (in cycads)—encircling of xylem of leaf traces from their beginning to their emerging into leaves

Globose—shaped like a ball or globe

Gum duct—canal along which secretion is frequently extruded

Gymnosperm—member of group of plants having exposed (naked) seeds in common (*see* p. 17)

Hardwood—deciduous angiospermous wood or plants

Hematite—(*see* p. 30)

Hemispheroidal—half-sphere shaped

Heterogenous ray—ray whose cells are unlike, usually consisting of upright and procumbent cells (*see* Fig. 227)

Heterosporous—said of plants having two types

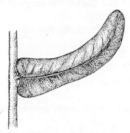

Figure 313. Fiber. a. Longitudinal section, showing pitting, b. Cross section

of spores: microspore (small, male) and megaspore (large, female)
Holdfast—rootlike structure of certain algae
Homogeneous ray—ray whose cells are all alike
(*see* Fig. 226)
Homosporous—said of plants having one type of
spore

Impervious—incapable of being penetrated
Impression—(*see* p. 28)
Index fossil—fossil characteristic of certain se-
quence of rocks; can be used to correlate
strata (rocks) for stratigraphic purposes
Indusium—(*see* p. 61)
Internodal—between nodes
Intersect—to divide into two parts by forming a
cross (*see* decussate)
Invertebrate—without backbone (e.g., brachio-
pod, clam, snail, etc.)
Isodiametric—having equal diameters

Lamina—thin, flat portion of leaf blade
Laminated—composed of thin layers or beds (*see*
p. 49)
Lanceolate—shaped like a spear point (*see* Fig.
329)
Leaf cushion—persistent rhomboidal or circular
area to which leaf was connected with stems
or branches in members of Lepidodendrales
(*see* p. 52)
Leaf scar—scar within leaf cushion where leaf
actually was attached (*see* p. 52)
Legume—pertaining to family Leguminosae
Ligule scar—(*see* p. 52)
Linear—narrow and having uniform width (*see*
Fig. 329)
Lobate (lobed)—composed of lobes (*see* Fig.
330)
Lobes—rounded division of structure, especially
of leaf (*see* Fig. 330)
Longitudinal—running lengthwise through stem
(e.g., In a stem, a longitudinal cut may parallel
the rays.)
Longitudinal or radial section—section formed
by cutting parallel to rays or along radial of
stem (*see* Fig. 314)

Marine—inhabiting, or formed from, sea
Megaspore—large spore that generally forms
female gametophyte (*see* p. 53)
Megaplant—visible fossil plant or plant remains
(as contrasted with microplant, pollen, spores,

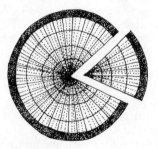

Figure 314. Radial cut

154

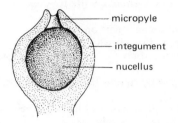

micropyle

integument

nucellus

Figure 315. Seed compression

Figure 316. Obovate shape

or microscopic plants)

Mesh—open spaces produced by net; spaces between veins in net-veined leaf (*see* Fig. 23)

Micropyle—opening between integuments through which pollen tube grows or pollen is drawn (*see* Fig. 315)

Microspore—small spore that usually forms male gametophyte

Monocot—informal abbreviation for monocotyledonae (*see* p. 21)

Montane—occurring on mountains

Moraine—mixed mass of rock deposited at end, or along sides, of glacier

Morphology—form, structure, and stages in the life of plants; division of botany that studies same

Multiple vessels—several vessels connected together in chain (*see* Fig. 267)

Multiseriate ray—ray composed of many rows of cells (*see* Fig. 227)

Net veined—(*see* Fig. 23)

Node—point on stem or branch at which one or more leaves are attached (*see* p. 55)

Nucellus—central portion of ovule, in which female gametophyte of seed plant develops (*see* Fig. 315)

Nutlet—small nut

Obcuneate—inverse wedge shaped (i.e., shaped like wedge but attached by broader end)

Obovate—egg shaped (ovate), attached by smaller end (*see* Figs. 316, 329)

Obtuse—rather blunt or roundish (applies to apex and base) (*see* Fig. 328); also angle of attachment of more than 90°

Orbicular—circular (*see* Fig. 329)

Order—(in plant and animal classification) subdivision larger than family but smaller than class

Ovate (ovoid)—shape like that of an egg cut longitudinally (*see* Fig. 329); broader end basal

Oxide—(*see* p. 30)

Palmately veined—(*see* Fig. 326)

Pantropical—growing or distributed throughout tropical regions

Parallel veined—(*see* Fig. 326)

Parasite—organism living in or on another organism and obtaining benefits from host, which it usually injures

Paratracheal wood parenchyma—parenchyma

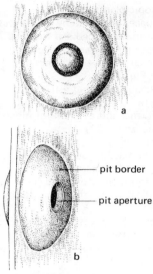

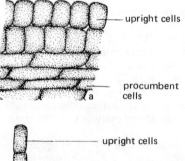

cells arranged around vessels

Parenchyma—isodiametric cells, usually with thin walls, that perform many life functions in a plant

Parichnos—two small scars on leaf scar of *Lepidodendrales* through which gases such as oxygen and carbon dioxide were exchanged (*see* p. 52)

Parted—divided by sinuses that extend almost to midrib

Peltate—attached in center of leaf or scale

Pendulous—hanging loosely

Perennial—living from year to year

Perforation plates—(*see* p. 138)

Pericycle—tissue that forms outermost layer of stele; in root, gives rise to secondary roots

Petiole—(*see* p. 95)

Petiolule—small petiole (e.g., petiole of leaflet)

Phloem—tissue that conducts food in plant

Photosynthesis—production of sugars by means of chlorophyll (*see* chlorophyll)

Phylloclades—flattened branch that performs function of leaf (e.g., photosynthesis)

Pinna (plur.: pinnae)—(*see* p. 60)

Pinnate—(*see* Fig. 327)

Pinnately compound—(*see* Fig. 327)

Pistillate cone—female cone

Pit—cavity in secondary wall (Point at which secondary wall overarches opening is called bordered pit.) (*see* Fig. 317)

Pith—tissue of generally thin-walled cells composing center of stems and some roots (*see* Fig. 219)

Plication—fold or plait

Polygon—figure usually having more than four sides, formed in leaf by anastomosing veins

Precipitate (n.)—substance separating from liquid (i.e., originally dissolved in liquid but, because of change in solution, became insoluble and dropped out)

Procumbent cell—horizontally elongated cell in ray; appears to be lying down (*see* Fig. 318)

Progymnosperms—group thought to be ancestors of gymnosperms

Propagate—to cause to continue or increase by sexual or asexual reproduction

Pteridosperm—group of extinct plants that resembled ferns but bore seeds (*see* p. 68)

Pumice—lightweight, porous volcanic rock

Pyriform—pear shaped (*see* p. 49)

Rachis—(*see* p. 60)

Rainshadow—the shadowing effect of mountains

pit border
pit aperture

Figure 317. Bordered pit. a. Longitudinal section, b. View showing both halves on cell wall

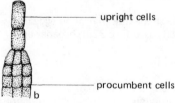

upright cells

procumbent cells

upright cells

procumbent cells

Figure 318. Ray. a. Radial section, b. Tangential section

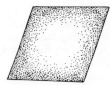

Figure 319. Rhombic shape

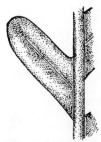

Figure 320. Sessile pinnule

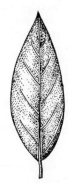

Figure 321. Simple leaf

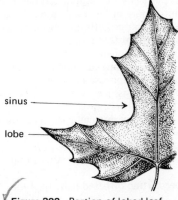

sinus ────────→

lobe ────

Figure 322. Portion of lobed leaf

that prevent moisture from moving inland and thereby produce inland deserts

Ray—(palm) plication in palm leaf that radiates from the rachis

Ray crossing—area of ray cell in contact with one tracheid

Reniform—kidney shaped

Reticulate venation (reticulation)—network pattern of veins in leaves (see p. 20)

Retuse—apex having notch in otherwise rounded tip

Rhizoids—hairlike projections from thallus that absorb water

Rhizome—underground stem (see p. 60)

Rhombic—oblique-angled figure resembling a diamond standing on one side (see Fig. 319)

Ring porous—(see p. 138)

Rounded—(see Fig. 328)

Rosette—cluster of leaves

Samara—dry, winged fruit

Saprophyte—plant that lives on dead organic materials (see p. 12)

Savanna—rather dry grasslands containing isolated trees

Scalariform perforation plates—(see p. 138)

Scrambler—plant that grows irregularly or climbs over some kind of support

Secondary thickening (or secondary wall)—innermost wall formed after cell ceased to elongate

Secondary venation—(see p. 95)

Septum (plur.: septa)—wall or partition

Serrate margin—(see Fig. 330)

Sessile—attached directly to base (see Fig. 320)

Shoot—branch or twig; ascending axis

Short shoots—short branches usually bearing leaves only (On ginkgo, however, they also bear fruiting structures.)

Silica—hard, glassy substance composed of silicon dioxide; occurs in many minerals, such as quartz and chalcedony

Simple leaf—undivided leaf (see Fig. 321)

Simple perforation plates—(see p. 138)

Sinus—notch between lobes in leaf (see Fig. 322)

Solenostelic—stele having phloem both inside and outside of xylem

Sorus (pl. sori)—cluster of plant reproductive bodies (see pp. 61, 67)

Species—in general, one of the lowest classifications in the plant kingdom (lowest is variety); name includes genus and species names (e.g.,

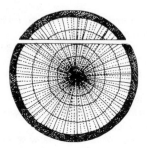

Figure 323. Tangential cut

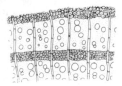

Figure 324. Terminal wood parenchyma

laminated algal mats

Abies concolor or white fir)
Spinose—covered with spines
Spiral—continuously circling around shoot axis
Spiral thickenings—secondary thickenings or
    walls arranged spirally within cell
Sporangium (plur.: sporangia)—structure in
    which spores are produced (*see* p. 53)
Sporangiophore—special branch or sporophyll
    that bears sporangia (*see* p. 15)
Sporophyll—modified leaf on which sporangia
    are borne (*see* p. 53)
Spur shoot—(*see* short shoots)
Stamen—pollen-bearing portion of flower
Staminate cone—male cone
Stele—vascular cylinder of plant, composed of
    xylem and phloem
Sterigma (plur.: sterigmata)—branch from which
    needles are abscissed, leaving persistent bases
Stimulus (plur.: stimuli)—mechanism, such as
    injury, that incites reaction or activity
Stoma (plur.: stomata)—small opening in epi-
    dermis of plant, surrounded with guard cells
Stomatal band—row of stomata
Stratigraphy—study of rock sequences
Stratum (plur.: strata)—layer of rocks deposited
    during given geologic period
Strobilus—cone
Stromatolites—(*see* p. 49)
Sustenance—support (e.g., food in plants)
Suture—line formed from fusion of margins; may
    also be line of weakness along which
    dehiscence occurs

Tangential section—cut made at right angle to
    rays (*see* Fig. 323)
Taxonomy—branch of biology concerned with
    naming and classifying plants and animals
Terminal wood parenchyma—aggregated paren-
    chyma forming more or less continuous layer
    at margin of season's growth (i.e., beginning
    or end of annual ring) (*see* Fig. 324)
Ternately arranged—arranged in threes
Tertiary venation—third-ranked veins (*see* p. 95)
Thallus—plant body of nonvascular plants; does
    not have definite roots, leaves, or stems (e.g.,
    algae, moss) (*see* p. 13)
Tracheid—elongated, tapering cell with secondary
    walls and perforations or pits on surface;
    used for support and water conduction
Trailing stem—specialized stem that runs along
    ground
Translucent—permitting diffused light to shine
    through

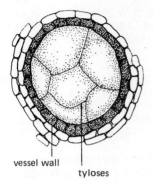

vessel wall

tyloses

Figure 325. Tyloses in vessel

Traumatic—resulting from violently produced wound or injury

Tripinnate—(*see* p. 61)

Triseriate—arranged in three rows

Truncate—apex or base of leaf having broad or squarish end (*see* Fig. 328, 331)

Tuff—(*see* p. 35)

Turbinate—cone shaped, with the smallest end down (*see* p. 49)

Two-ranked—arranged in two vertical rows on opposite sides of shoot or stem

Tylose—portion of cell protruding into vessel; in cross section, appear as a group of thin lines (*see* Fig. 325)

Umbo—diamond shaped structure or boss in middle of apophysis on pine cones (*see* p. 90)

Understory plant—plant that grows beneath canopy of higher trees

Undulate—having wavy margin (*see* Fig. 330)

Uniseriate ray—ray consisting of single row of cells (*see* Fig. 226)

Unstoried (nonstoried) rays—rays not aligned when viewed tangentially

Upright cell—vertically elongated cell in a heterogeneous ray (*see* Fig. 318) (cf. procumbent cell)

Valve—one of fragments into which capsule will disintegrate at maturity

Vascular—said of plant having conductive tissue

Vascular bundle—(*see* bundle)

Vessel—tubelike water conductor in xylem of angiosperm and, infrequently, of other groups

Volcanic dust—dust-sized rock particles blown from volcanoes

Whorl—(*see* p. 55)

Selected Readings

Alvin, K. L., P. D. W. Barnard, and W. G. Chaloner, eds. 1968. *Studies on fossil plants.* London: Academic Press.

Andrews, H. N. Jr. 1947. *Ancient plants and the world they lived in.* Ithaca, New York: Comstock Publishing Associates.

_____. 1961. *Studies in paleobotany.* New York: John Wiley and Sons.

Arnold, C. A. 1947. *An introduction to paleobotany.* New York: McGraw-Hill Book Co.

Banks, H. P. 1970. *Evolution and plants of the past.* Belmont, California: Wadsworth Publishing Co.

Becker, H. F. 1961. Oligocene plants from the Upper River Basin in southwestern Montana. *Geol. Soc. Amer. Mem.* 82:1-127.

_____. 1969. Fossil plants of the Tertiary Beaverhead Basins in southwestern Montana. *Palaeontographica* 127B:1-142.

Darrah, W. C. 1960. *Principles of paleobotany.* New York: John Wiley and Sons.

Dunbar, C. O. 1960. *Historical geology.* New York: John Wiley and Sons.

Graham, A. 1965. The Sucker Creek and Trout Creek floras of southeastern Oregon. *Kent State Univ. Bull., Res. Ser.* 9:1-147.

Greguss, P. 1967. *Fossil gymnosperm woods in Hungary from the Permian to the Pliocene.* Budapest: Akademial Kiado.

_____. 1969. *Tertiary angiosperm woods in Hungary.* Budapest: Akademial Kiado.

Johnson, J. H. 1969. *Limestone-building algae and algal limestones.* Boulder, Colorado: Johnson Publishing Co.

MacGinitie, H. D. 1941. A Middle Eocene flora from the central Sierra Nevada. *Carnegie Inst., Wash. Publ.* 534:1-178.

_____ . 1953. The fossil plants of the Florissant beds. *Carnegie Inst., Wash. Publ.* 599:1-198.

Muller, W. H. 1963. *Botany: a functional approach.* New York: MacMillan Co.

Ransom, J. E. 1964. *Fossils in America.* New York: Harper and Row.

Rushforth, S. R. 1971. A flora from the Dakota Sandstone Formation (Cenomanian) near Westwater, Grand County, Utah. *Brigham Young Univ. Biol. Ser.* 14:1-44.

_____ , and W. D. Tidwell. 1972. *Plants and man.* Minneapolis: Burgess Publishing Co.

Seward, A. C. 1898-1919. *Fossil plants* I-IV. New York: Hafner Publishing Co. (Reprinted 1963).

Tidwell, W. D. 1967. Flora of Manning Canyon Shale. Part I: a lowermost Pennsylvanian flora from the Manning Canyon Shale, Utah, and its stratigraphic significance. *Brigham Young Univ. Geol. Studies* 14:3-66.

Tidwell, W. D., et al. 1970. *Plants through time: laboratory manual for introductory paleobotany.* Dubuque, Iowa: Wm. C. Brown Company Publishers.

_____ , D. A. Medlyn, and G. F. Thayn. 1972. Fossil palm materials from the Tertiary Dipping Vat Formation of central Utah. *Great Basin Nat.* 32:1-15.

_____ , D. A. Medlyn, and A. D. Simper. 1974. Flora of Manning Canyon Shale. Part II: Lepidodendrales. *Brigham Young Univ. Geol. Studies* 21(3):119-46.

White, D. 1929. Flora of the Hermit Shale, Grand Canyon, Arizona. *Carnegie Inst., Wash. Publ.* 405:1-221.

Wilson, C. L., W. E. Loomis, and T. A. Steeves. 1971. *Botany.* 5th ed. New York: Holt, Rinehart and Winston.

Woodford, A. O. 1965. *Historical geology.* San Francisco: Freeman and Co.

Outline Key

Ferns and Fernlike Foliage

Gleichenia, p. 63

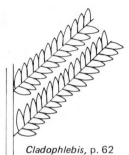

Cladophlebis, p. 62

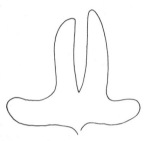

Sphenopteridium, p. 70
Sphenopteris, p. 69

Clathropteris, p. 65
Lygodium (climbing fern), p. 63

Lygodium (climbing fern), p. 63

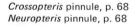

Crossopteris pinnule, p. 68
Neuropteris pinnule, p. 68

Wingatea, p. 64

Crossopteris, p. 68
Neuropteris, p. 68
Pecopteris, p. 70

Alloiopteris, p. 62
Corynepteris, p. 62

Cynepteris, p. 63
Gleichenia, p. 63

Coniopteris fertile
pinnae, p. 64

Malatonidium, p. 65
Phlebopteris, p. 65

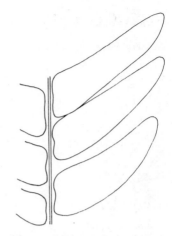

Pecopteris, p. 70

Neuropteris, p. 68

Astralopteris, p. 66

Osmunda, p. 62

Asplenium (spleenwort), p. 66
Crossopteris, p. 68
Odontopteris, p. 69

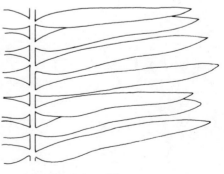

Astralopteris, p. 66

Rhodea, p. 69
Sphenopteris, p. 69
Zeilleria, p. 70

Rhodea, p. 69
Zeilleria, p. 70

Sphenopteris, p. 69

Seeds, Fruits, and Microsporangiate and Seedlike Structures

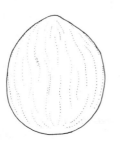

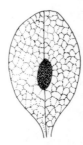

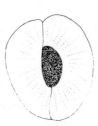

Juglans (walnut), p. 113

Koelreuteria (goldenrain tree), p. 102

Ptelea (hop tree), Ulmus (elm), p. 104

Rigbyocarpus, p. 72

Cornucarpus, p. 72

Engelhardtia, p. 114

 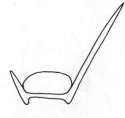  

Cardiocarpus, p. 71   *Lepidocarpon,* p. 54   *Gnetopsis,* p. 71   *Jensensispermum,* p. 74

 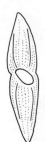 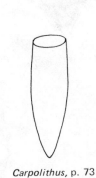 

Behuninia, p. 74   Legume pod, p. 106   *Carpolithus,* p. 73   *Juglans* (walnut), p. 113

Ailanthus (tree of heaven), p. 99   *Equisetum* (horsetail) bulbils, p. 58   *Aulacotheca,* p. 72

   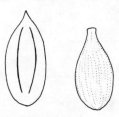

Lomatia, p. 108   *Carpolithus,* p. 73   *Trigonocarpus,* p. 71

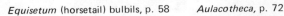

Coniferophyta

Pinus (pine), p. 90   *Abies* (fir), p. 90   *Picea* (spruce), p. 92   *Tsuga* (hemlock), p. 91

Araucaria (Norfolk Island pine), p. 85   *Glyptostrobus* (water pine), p. 88   *Cordaianthus*, p. 84

Sequoia (redwood), p. 88   *Metasequoia* (dawn redwood), p. 89   *Chamaecyparis* (white cedar), p. 8

Pinus (pine) seed, p. 90   *Picea* (spruce) seed, p. 92   *Tsuga* (hemlock) seed, p. 92   *Abies* (fir) seed, p. 91

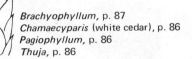

Araucaria (Norfolk Island pine), p. 85
Sequoia (redwood), p. 88
Walchia, p. 84

Brachyophyllum, p. 87
Chamaecyparis (white cedar), p. 86
Pagiophyllum, p. 86
Thuja, p. 86

Metasequoia (dawn redwood), p. 89
Sequoia (redwood), p. 88
Taxodium (bald cypress), p. 88

Podozamites, p. 94

Abies (fir) cone scale, p. 91   Picea (spruce) cone scale, p. 92   Cordaites, p. 83

Araucaria (Norfolk Island pine)
cone scale, p. 85

Abies (fir) needle, p. 90

Pinus (pine), p. 90   Sequoia (redwood) seed, p. 88   Protophyllocladus, p. 93

Lobed Leaves

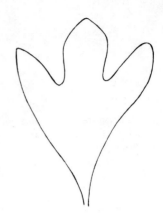

Acer (maple), p. 103
Cardiospermum (balloon vine), p. 102
Platanus (sycamore), p. 108

Sassafras, p. 97

Acer (maple), p. 103
Cardiospermum (balloon vine), p. 102
Platanus (sycamore), p. 108

Platanophyllum, p. 108

Quercus (oak), p. 112

Leaves and Leaflets

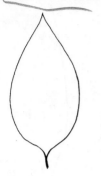

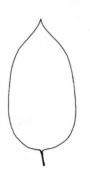

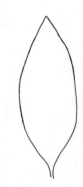

Betula (birch), p. 110
Populus (poplar), p. 109

Betula (birch), p. 110
Fagopsis, p. 111

Caesalpinites leaflet, p. 106
Carpinus (hornbeam), p. 112
Cinnamomum, p. 97
Crataegus (hawthorn), p. 107
Fagopsis, p. 111
Magnolia, p. 96
Quercus (oak), p. 112
Rhus (sumac) leaflet, p. 101
Robinia (locust) leaflet, p. 106
Sassafras, p. 97
Zelkova (keaki tree), p. 115

Ginkgo (Maidenhair tree), p. 81

Cercocarpus (mountain mahogany), p. 107

Castanea (chestnut), p. 113
Dryophyllum, p. 113
Ficus, p. 104
Mimosites leaflet, p. 105
Populus (poplar), p. 109
Quercus (oak), p. 112
Sapindus (soapberry) leaflet, p. 101

Amelanchier (serviceberry), p. 106
Betula (birch), p. 110
Carpinus (hornbeam), p. 112
Carya (hickory), p. 114
Cinnamomum, p. 97
Ficus, p. 104
Juglans (walnut) leaflet, p. 113

Typha (cattail), p. 115

169

Cedrela leaflet, p. 100
Sapindus (soapberry) leaflet, p. 101

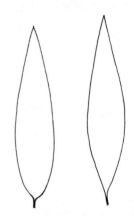

Pterocarya (wingnut), p. 114
Rhus (sumac) leaflets, p. 101
Salix (willow), p. 110

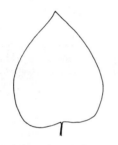

Alnus (alder), p. 110
Cercidiphyllum (katsura tree), p. 97
Ficus, p. 104
Populus (poplar), p. 109
Robinia (locust) leaflet, p. 106

Ulmus (elm), p. 104

Lobed and Compound Leaves

Oreopanax, p. 109

Mahonia (Oregon grape) leaflet, p. 98

Acer (maple, p. 103
Ptelea (hop tree), p. 99

Caesalpinites, p. 106
Carya (hickory), p. 114
Cedrela, p. 100
Juglans (walnut), p. 113
Koelreuteria (goldenrain tree), p. 102
Lomatia, p. 108
Mimosites, p. 105
Rhus (sumac), p. 101
Robinia (locust), p. 106
Sapindus (soapberry), p. 101

Fruits, Flowers, and Seeds

Fagopsis, p. 111

Carya (hickory), p. 114

Astronium, p. 100

Ficus, p. 73

Acer (maple) samara, p. 103

Fagopsis immature fruit, p. 111

Cercocarpus (mountain mahogany), p. 107

Ficus, p. 73

171

Calamitean Foliage

Annularia, p. 56

Asterophyllites, p. 55

Cycadophyta Foliage

Pseudocycas, p. 78

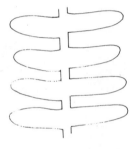

Ctenis, p. 75
Otozamites, p. 79
Pseudoctenis, p. 76
Pterophyllum, p.78
Ptilophyllum, p. 77
Zamites, p. 77

Nilssonia, p. 75
Otozamites, p. 79

Lepidophloios, p. 53

Lepidodendron, p. 51

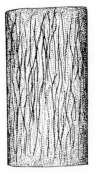

Tempskya, p. 125

Cycadeoidea, p. 127

Sigillaria, p. 52

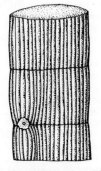

Archaeocalamites, p. 56
Calamites, p. 55
Mesocalamites, p. 56

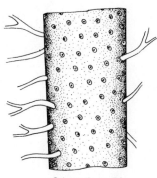

Stigmaria, p. 54

Index

Abies (fir) 45, 90
Abietites 91
Acer (maple) 43, 44, 45, 47, 103, 137, 142
Aceraceae 103, 142
Acuminate 94, 96, 98, 100, 101, 102, 104, 109, 110,
    111, 112, 113, 114
Acuminate apex 78, 96, 98, 100, 101, 102, 104, 109,
    110, 111, 112, 113, 114 (*see* Fig. 328)
Acute 86, 88, 92, 97, 98, 100, 102, 106, 107, 108, 109,
    110, 111, 113, 114, 115
Acute apex 64, 76, 77, 79, 86, 88, 92, 96, 97, 98, 100,
    102, 106, 107, 108, 109, 110, 111, 113, 114, 115
    (*see* Fig. 328)
Acute base 107 (*see* Fig. 331)
Adiantum (maidenhair fern) 18, 44
Aerenchyma tissue 51
Agathis 94
Ailanthus (tree of heaven) 45, 99
Alberta 22
    Dunvegan flora 23
    Edmonton Formation 23, 32
    Frenchman Formation 23, 32
    Kootenay Formation 23, 32
    Milk River flora 23
    Paskapoo Formation 23, 32
Alder (*see Alnus*)
Algae 11, 12, 24, 35, 42, 49
Algal balls 50
Algal fungi 13
Algal reef 49
Allantodiopsis 42, 66
Allantodiopsis erosa 66
Alloiopteris 36, 62
Alnus (alder) 44, 110
Altamont Pass, California 32
Amber 27
Amelanchier (serviceberry) 44, 106
Anacardiaceae 100, 141
Anacardioxylon 141
Anatomy 119
Andrews, Oregon 31
Angiosperm wood 122
Angiosperms 13, 17, 18, 19, 25, 41, 42, 43, 46, 75, 95,
    119, 137
Annual rings 129
Annularia 36, 39, 55
Anthophyta (Angiosperms)
Apex (plur.: apices) (*see* Fig. 328)
Apophyses 90
Araliaceae 109
Araucaria (Norfolk Island pine) 42, 85
Araucariaceae 85, 130
Araucarioxylon 39, 40, 130
Arbutus 44
Arborvitae 18
Archaeocalamites 36, 56
Arctocarpus 43
Arizona
    Chinle Formation 33, 40, 58, 62, 63, 64, 65, 130,

131, 136
Dakota Formation 67
Petrified Forest National Park 29
Artemesia tridentata (sagebrush) 46
Artisia 36, 83
Ash (*see Fraxinus*)
Asplenium 42, 66
Asterophyllites 36, 38, 55
Astralopteris 41, 66
Astronium 100, 101
Aulacotheca 36, 72
Autophycus 35

Bacteria 13
Bald cypress (*see Taxodium*)
Balloon vine (*see Cardiospermum*)
Bark 120
Bars of Sanio (crassulae) 129, 131, 133, 135
Bases (*see* Fig. 331)
Beartooth Butte flora 23, 35
Beaverhead flora 23, 106, 109
Beech (*see Fagus*)
Behuninia 40, 74
Bellevue, Alberta 32
Belt Series 49
Berberidaceae 98
Betula (birch) 45, 110, 111, 142
Betulaceae 110
Big Horn Mountains 50
Binomial nomenclature 21
Birch (*see Betula*)
Biseriate pitting 129
Biseriate ray 132, 140, 141, 145
Biserrate 110
Blackhawk Formation 23, 42
Black Hills 10
Blade 59, 95
Blairmore, Alberta 32
Blue Forest, Wyoming 32
Blue green algae (*see* Cyanophyta)
Blunt apex 65, 87, 94, 107, (*see* Fig. 328)
Bogs 13
Bolster 52, 53
Bonanza, Utah 32
Bordered pit 123
Brachyphyllum 86, 87
Bract 52, 56
Bread mold (*see Rhizopus*)
Bridge Creek flora, 23, 32, 115
British Columbia 22
    Dunvegan flora 23
    Jackass Mountain Group 23, 32
    Nanaimo Group 23, 32
    Smithers flora 23, 63
    Spence Bridge Group 23, 33
Broad-leafed tree 21
Broad rays 146
Broggeria 35
Brongniart 9

Bryophyta 11, 12, 13, 42
Bulbil 58
Bundle cap 148
Burns, Oregon 32
Burro Canyon Formation 41

Caesalpinites 106
Calamitales 37, 41, 55
Calamites 15, 24, 36, 38, 39, 55, 56
Calamostachys 56, 57
California
    Chalk Bluffs flora 23, 32
    LaPorte flora 23
    Mount Eden Beds 31
    San Pablo flora 23, 32
    Weaverville flora 23, 32
Cambrian 35
Capparidaceae 140
Carboniferous 14, 15, 16, 17, 38, 50, 55, 58, 59, 68, 69
Cardiocarpus 36, 71
Cardiospermum (balloon vine) 44, 102
Cardiospermum halicacubum 102
Carpinus 112
Carpolithus 73
Carya (hickory) 114,
Castanea (chestnut) 113
Castle Dale, Utah 32
Casts 28, 54
C-shaped petiole 124
"Catchall" taxa 104
Cedar Mountain Formation 23, 32, 40, 41, 139
Cedrela 100
Cedrela mexicana (mexican cedar) 100
Cedroxylon 135
Celtis 44, 45
Cenozoic 35, 42
Cephalotaxaceae 94
Cephalotaxus 40, 44, 94
Cephalotaxus nevadensis 94
Cercidiphyllaceae 97
Cercidiphyllum 42, 97
Cercocarpus 107
Chalk Bluffs flora 23, 32, 63, 100, 101, 108
Chamaecyparis 44, 86
Chestnut (*see Castanea*)
China 89, 114
Chinle Formation 33, 40, 58, 62, 63, 64, 65, 130, 131, 136
Chlorellopsis 50
Chlorophyta (green algae) 12, 50
Chrysophyta (golden brown algae) 12
Cinnamomum 42, 97
Cirque 47
Cladophlebis 40, 62
Clarno flora 23, 44, 108, 125
Classes 21
Clathropteris 40, 65
Clayton, California 32

Climatic zone 38
Climbing fern (see *Lygodium*)
Club moss (see Lycopodophyta)
Coal 16, 42
Coal balls 27
Coal Flora of Pennsylvania 10
Codonophycus 50
Collenia 35, 49
Colorado
  Burro Canyon Formation 41
  Dolores Formation 58
  Florissant flora 32, 67, 99, 101, 102, 106, 109, 111,
    112, 114, 115
  Mosquito Range flora 23
  Salida Canyon flora 23
Colorado Springs 44
Comax, British Columbia 32
Common sumac (see *Rhus glabra*)
Compound leaf 59, 96, 100, 101, 102, 103, 106
Compression 27, 54
Comstock flora 23
Cones 39
  Angiosperm 111
  Calamites (Calamostachys) 56, 57
  Calamites (Palaeostachya) 56, 57
  Conifer 82, 83, 85, 87, 88, 89, 90, 91, 92
  Cordaites (Cordaianthus) 84
  Lepidodendron (Lepidostrobus) 52, 53
Cone scales 85, 91, 92
Coniferales 13, 37, 41, 82, 84
Coniferales incertae sedis 94, 134
Coniferophyta (conifers) 13, 17, 18, 24, 39, 42, 44, 82,
  129
Coniopteris 40, 42, 64
Copper Basin flora 23, 32, 111
Cordaianthus 36, 38, 84
Cordaitales 13, 37, 83
Cordiates 24, 36, 38, 83, 84
Cordate (heart-shaped) leaf 63, 96, 109
  (see Fig. 329)
Cordate base 98, 101, 110 (see Fig. 331)
Cornucarpus 36, 72
Cornucarpus longicaudatus 72
Cortex 119, 124, 125
Corynepteris 36, 38, 62
Cotyledon 21
Crassulae (see Bars of Sanio)
Crataegus (hawthorn) 44, 107
Creede flora 23
Crenate margin 67, 96, 100, 109 (see Fig. 330)
Crenate-dentate 100
Crenate-serrate 100
Cretaceous 25, 31, 39, 41, 42, 62, 63, 64, 65, 66, 67,
  73, 76, 77, 78, 79, 85, 87, 88, 89, 91, 93, 95, 97,
  103, 104, 109, 110, 113, 115, 124, 125, 127, 134,
  139, 142
Crossopteris 36, 38, 68
Crossover field 123, 130
Cross section 117, 118, 119

Cryptozoon 49
Ctenis 40, 75
Cuneate (wedge shaped) base 69, 70, 92, 93, 98, 101,
    102, 104, 105, 106, 107, 108, 109, 110, 111, 113,
    114, 115 (*see* Fig. 331)
Cunninghamia 94
Cupressaceae 86
Cupressinoxylon 132, 134
Cuticle 28, 75
Cyanophyta (blue green algae) 12 49
Cycadales 12, 41, 74, 75, 127
Cycadeoidea 12, 40, 127
Cycadeoidales 12, 41, 77, 127
Cycadeoids 39, 94
Cycadophyta (cycads) 12, 17, 24, 42, 44, 75, 127
Cynepteris 40, 63

Dadoxylon 130
Dakota Formation 23, 32, 63, 64, 65, 67
Da Vinci 9
Dawn redwood (*see Metasequoia*)
Dawson 10
Dayville, Oregon 32
Deltoid 63, 87, 97 (*see* Fig. 329)
Deltoid-ovate 98
Dendrite 30
Dentate margin 65, 93, 96, 97, 98, 101, 103, 106, 110,
    111, 112, 113, 115 (*see* Fig. 330)
Deschutes Formation 23, 31
Devonian 24, 35, 59
Diaphragms 83
Dicksoniaceae 64
Dicotyledonae (dicots) 13, 20, 21, 96, 137
Diffuse-porous wood 137, 139, 140, 141, 142, 144,
    145, 146, 147
Dinosaurs 39
Dipteridaceae 65
Distinct growth rings 133, 135, 141, 142, 143, 144,
    146, 147
Division 24
Dolores Formation 58
Doubly serrate 102, 104, 111, 112 (*see* Fig. 330)
Douglas fir (*see Pseudotsuga*)
Drumheller, Alberta 32
Dryophyllum 42, 113
Dryophyllum subfalcatum 113
Dryopteris 67
Dunvegan flora 23

Eastend, Saskatchewan 32
Eden Valley, Wyoming 32, 50, 140, 141, 145, 148
Edenoxylon 141
Edmonton flora 23, 32
Elators 15
Elko, Nevada 88
Ellensburg flora 23
Elliptical 64, 78, 98, 107, 109 (*see* Fig. 329)
Elm (*see Ulmus*)
Elsinore, California 31

Emarginate apex 99, 106 (*see* Fig. 328)
Endodermis 124, 126, 148, 149
Engelhardtia 114
Entire margin 93, 96, 97, 98, 100, 101, 103, 104, 106,
    108, 109, 110, 112, 113 (*see* Fig. 330)
Eocene 43, 44, 50, 98, 100, 114, 115
Ephedra (Mormon tea) 13, 19
Epidermis 120, 126
Epoxy resin 117
Equisetales 58
Equisetites 58
Equisetum (horsetail, scouring rush) 12, 15, 19, 40, 55,
    58, 73
Eucalyptus 42
Euphorbiaceae 140
Evanston, Wyoming 32
Evergreens 82

Fagaceae 111, 146
Fagara 141
Fagopsis 45, 111
Fagus (beech) 45, 146
Falcate (sickle shaped) leaf 62, 77, 86, 105
False trunk 125, 126
Families 21, 24
Farson, Wyoming 32
Fascicle 90
Ferns (*see* Filicophyta)
Fern allies 12, 14, 42
Fernlike foliage 37, 38, 41, 68
Fiber tracheids 120, 121
Ficus (fig) 42, 44, 73, 104
Fiddlehead 59
Fig (*see Ficus*)
Filicophyta (ferns) 12, 14, 16, 24, 25, 26, 39, 41, 42,
    44, 59, 60, 67, 120, 124
Fingernail polish 28
Fingerrock flora 23, 45
Fir (*see Abies*)
Flagstaff Limestone Formation 50
Floral parts 21
Florissant Beds National Monument 32
Florissant, Colorado 32
Florissant flora 23, 32, 43, 67, 99, 100, 101, 102, 106,
    109, 111, 112, 114, 115
Flowering plants (see angiosperms)
Fontanelle, Wyoming 32
Forchhammeria 140
Forchhammerioxylon 140
Form genus 24, 55, 71
Form species 24
Fort Union Formation 23, 32, 42, 43, 124
Fossil, Oregon 32
Fossil wood 117
Foxhill Sandstone 73
Fraxinus (ash) 43, 45
Frenchman Formation 23, 32
Frond 59, 60, 124
Frontier Formation 67

Fruit 19, 73, 99, 100, 102, 103, 104, 108, 111, 114
Fungi 11, 12, 13, 42
Fusiform 99

Genomites 42, 115
Genus 21
Giant redwoods (see Sequoiadendron)
Ginkgo 12, 42, 81
Ginkgo biloba 81
Ginkgo State Park 144
Ginkgophyta (ginkgos) 12, 17, 18, 24, 42, 81
Glacier National Park 35, 49
Glass slide 117
Gleichenia 40, 41, 63
Gleicheniaceae 63
Glyptostrobus 88
Gnetales 19, 136
Gnetophyta 13, 17, 19
Gnetopsis 36, 71
Gnetum 19
Goldenrain tree (see Koelreuteria)
Grand Canyon 39, 84
Grand Canyon National Park 33
Grass Valley, California 32
Grasses 21
Green River Formation 23, 32, 43, 100, 101, 102, 105, 106, 109, 111, 114, 115
Green River System 43
Greybull, Wyoming 33
Grinding wheel 117
Grit 117, 119
Ground pine (see Lycopodium)
Growth rings 132, 133, 135, 139, 140, 141, 142, 143, 144, 146, 147
Gum ducts 141, 142
Gymnosperms 12, 17, 39, 82

Hadrophycus 35
Hanksville, Utah 33
Hawthorn (see Crataegus)
Heart shaped (see cordate)
Hematite 30
Hemlock (see Coniferophyta)
Hepaticae 13
Herbarium Diluvianum 9
Hermit Shale Formation 23, 33, 84
Heterogeneous rays 122, 123, 138, 139, 140, 141, 145, 147
Hevea braziliensis (rubber tree) 140
Hevea microphylla 140
Heveoxylon 140
Hickory (see Carya)
Holdfast 11
Homogeneous rays 122, 123, 138, 142, 143, 144, 147
Homosporous 14, 15
Honaker Trail flora 23, 36, 70
Hooke 9
Hop hornbeam (see Ostrya)
Hop tree (see Ptelea)

Horsetail (*see Equisetum*)
Horseshoe-shaped xylem 124
Hostinella 35

Icasinoxylon 41
Idaho
    Latah Formation 23, 31, 45, 88, 100, 114, 115
    Payette, flora 111, 114
    Salmon flora 23, 32
    Sucker Creek flora 23, 32, 45, 99, 100, 109, 114, 115
    Thorn Creek flora 23, 45, 111, 114
Ilex 42
Impressions 27
India 114
Indistinct growth rings 140, 141, 145
Indusium 60
Internal structure 119
Isoetales 14
Isoetes 14, 51

Jackass Mountain Group 23, 32
Jarbidge, Nevada 44
Jensensispermum 40, 74
Joint grass (*see* Sphenophyta)
Juglandaceae 113, 147
Juglans (walnut) 113, 147
Juniper 86
Jurassic 39, 41, 64, 65, 74, 77, 78, 86, 93, 124, 130, 135

Keteleeria 94
Koelreuteria (goldenrain tree) 45, 102
Kootenay Formation 32, 40
Korea 89

Lake Bonneville 48
Lake Erie 48
Lake Lahontan 48
Lake Michigan 48
Lake Mountain, Utah 33
Lake Ontario 48
Lakeside "70" 117
Lanceolate leaf 62, 64, 66, 75, 77, 85, 94, 96, 100, 104, 110, 114 (*see* Fig. 329)
LaPorte flora 23
Larch (*see Larix*)
Large rays 140, 146
Large tyloses 145
Larix (larch) 82
Latah Formation 23, 31, 45, 88, 100, 114, 115
Lauraceae 97, 139
Laurel (*see Laurus*)
Laurus (laurel) 42, 44
Leaf bases 124
Leaf cushions 51, 52
Leaf gap 120
Leaf margins (*see* Fig. 330)
Leaf scar 52

Leaf trace 120
Leaves
  Angiosperms 95-116
  Calamites 55, 56
  Conifers 18, 84-94
  Cordaites 84
  Cycads 75-79
  Ferns 59, 60-67
  Fernlike foliage 39, 68-70
  Ginkgos 81
  Lycopodophyta 51
Lebachiaceae 84
Legume 106
Leguminosae 105, 144
Lehi, Utah 33
Lepidocarpon 36, 54
Lepidodendrales 37, 51
Lepidodendron 12, 24, 28, 36, 38, 51, 52, 54
Lepidophylloides 24, 37, 51
Lepidophloios 37, 54
Lepidostrobophyllum 37, 54
Lepidostrobus 24, 37, 54
Lesquereux 10
Lewiston, Idaho 31
Lilies 21
Linear leaf 75, 77, 78, 90, 96, 115 (*see* Fig. 329)
Linear lanceolate leaf   65, 66, 75 (*see* Fig. 329)
Liriodendron 95
Lithocarpus (oak) 44
Livermore, California 32
Liverworts 13
Lobate (lobed) margins 67, 97, 102, 107, 108, 109 (*see* Fig. 330)
Locust (*see Robinia*)
Lomatia 45, 108
Luther 9
Lycopodiales 14
Lycopodium (ground pine) 14, 51
Lycopodophyta (club moss) 12, 14, 16, 51
Lycopods 24, 39
Lygodium (climbing fern) 63
Lytton, British Columbia 32

Magnolia 42, 96
Magnoliaceae 96
Mahonia (Oregon grape) 44, 45, 98
Maidenhair fern (*see Adiantum*)
Manning Canyon Shale 23, 33, 36, 38, 55, 56, 68, 70, 71, 72
Maple (*see Acer*)
Mascall flora 23, 32, 45, 88, 114, 115
Matoniaceae 65
Matonidium 41, 65
Medicine Bow flora 23
Medicine Bow Mountains, Wyoming 35, 42
Megaspore 14, 52
Meliaceae 100
Menispermaceae (moonseed family) 98
Menispermites 42, 98

Mesa Verde Group 32
Mesocalamites 37, 56
Mesozoic 24, 35, 39, 40, 59, 75, 78, 133
Metasequoia (dawn redwood) 42, 45, 82, 89
Metasequoia occidentalis 89
Mexican cedar (*see Cedrela mexicana*)
Microsporangiate structures 37, 71
Microsporangium 89
Microspore 14, 52
Midveins 95
Migrations 42
Milk River flora 23
Mimosites 44, 105
Miocene 42, 45, 46, 47, 100, 111, 143, 144, 146, 147
Mississippian 24, 38, 50, 70
Moab, Utah 32, 33, 39, 128
Molds 13, 28
Monanthesia 40, 128
Monocotyledonae (monocots) 20, 21, 115, 147
Montana
    Beaverhead flora 23, 106, 109
    Fort Union Formation 23, 32, 42, 43, 124
    Kootenay Formation 23, 32, 40
    Morrison Formation 23, 33, 39, 40, 73, 74, 134, 135
    Ruby flora 23, 32, 45
    York Ranch flora 23
Moonseed family (*see Menispermaceae*)
Moraceae 104
Mormon tea (*see Ephedra*)
Morrison Formation 23, 33, 39, 40, 73, 74, 130, 134,
    135
Mosquito Range flora 23, 36
Mosses (*see Musci*)
Mount Eden Beds 23, 31
Multiseriate rays 123, 130, 138, 139, 144
Musci (mosses) 13
Mushrooms 13
Myrica 145
Myricaceae 145

Nanaimo, British Columbia 32
Nanaimo Group 23, 32
Narrow rays 146
Needles 84, 85, 89, 90, 92, 94
Neocalamites 40, 56
Neuropteridaceae 68
Neuropteris 37, 38, 68, 71
Nevada
    Copper Basin flora 23, 32, 111
    Elko 88
    Fingerrock flora 23, 45
Nevada City, California 144
New Mexico
    Chinle Formation 33
New Zealand 93
Nicols 9
Nilssonia 40, 75
Nilssoniales 75
Nonporous 120

Nonray 122
Nonseptate fibers 141, 144
Nonseptate fiber tracheids 144
Nonstoried rays 142, 144, 147
Norfolk Island pine (see *Araucaria*)
North Dakota
    Fort Union Formation 23, 32, 42, 43, 124
    Foxhill Sandstone 73

Oak (see *Lithocarpus, Quercus*)
Oblanceolate leaf 107 (see Fig. 329)
Oblong leaf 63, 96, 98 (see Fig. 329)
Oblong-lanceolate leaf 93, 113
Oblong-obovate leaf 93
Oblong-ovate leaf 111, 112, 115
Obovate leaf 96, 97, 104, 106
Obtuse apex 64, 76, 79, 85, 92, 93, 94, 104, 105 (see
    Fig. 328)
Obtuse base 107 (see Fig. 331)
Oculipores 130
Odontopteris 37, 39, 69
Oligocene 44, 86, 88, 99, 100, 112, 147
Oopores 130
Opportunity, Washington 31
Orbicular (round) leaf 96 (see Fig. 329)
Order 21, 24
Oregon
    Bridge Creek flora 23, 32, 115
    Clarno Formation 23, 108, 125
    Comstock flora 23
    Deschutes Formation 23, 31
    Mascall Formation 23, 32, 45, 88, 114, 115
    Spotted Ridge flora 23, 55, 56, 70
    Stinking Water flora 23, 32, 45, 99, 115
    Sucker Creek flora 23, 32, 45, 99, 100, 109, 114,
    115
    Trout Creek flora 23, 31, 45, 99, 115
Oregon grape (see *Mahonia*)
Oreopanax 109
Osmunda 42, 44, 62, 124
Osmunda braziliensis 124
Osmunda carnieri 124
Osmundacaulis 8, 39, 40, 124
Osmundaceae 62
Ostrya (hop hornbeam) 111
Otozamites 79
Oval leaf 96, 97, 100, 104, 108 (see Fig. 329)
Ovary 19
Ovate leaf 63, 96, 99, 100, 102, 104, 106, 108, 109,
    110, 111, 112, 114 (see Fig. 329)
Ovate-elliptic leaf 97
Ovate-lanceolate leaf 101, 108, 109, 113
Ovate-rhombic leaf 106

Pagiophyllum 86
Painted Hills State Park 32
Paleostachya 37, 56
Paleocene 32, 42, 66, 89, 97, 115, 124
Paleozoic 24, 35, 38, 59

Palmae (palms) 115, 148
Palmate shaped or lobed 63, 98, 103, 108, 109 (see Fig. 329)
Palmate veined 96, 97, 98, 103, 105 (see Fig. 326)
Palmocarpon 73
Palmoxylon 148
Palms (*see Palmae*)
Paraphyllanthoxylon 40, 41, 139
Paraphyllanthoxylon arizonense 139
Paraphyllanthoxylon idahoense 139
Paraphyllanthoxylon utahense 139
Parasite 12
Parenchyma 120, 121, 122, 129, 131, 132, 135, 138, 139, 140, 141, 142, 143, 144, 145, 146, 147
Parichnos 51
Parsons 9
Paskapoo Formation 23, 32
Payette flora 111, 114
Peat 13
Pecopteridaceae 70
Pecopteris 37, 70
Peltate scales 86, 88
Pennsylvanian 24, 38, 39, 71
Peppertree (*see Schinus*)
Perforation plates 121, 138
Periderm 119, 120, 122
Permian 39, 68, 71
Persea 108
Petiole 59, 75, 95
Petrifaction 27, 29, 117
    Angiosperms 137-149
    Conifers 129-136
    Cycads 127-128
    Ferns 124-126
Petrified Forest National Park 23, 29, 33, 39 (see also Chinle Formation)
Phaeophyta (brown algae) 12
Phlebopteris 40, 65
Phloem 119, 120, 121, 122, 125, 126, 148, 149
Phoenicites 42, 115
Phylloclad 93
Phyllocladus 93
Picea (spruce) 18, 45, 92
Pinaceae 90, 133
Pine (*see Pinus*)
Pine seeds 90
Pinnae 59, 60, 61, 64, 65, 66, 67, 68, 75, 76, 77, 78, 79, 80
Pinnate 59, 60, 63, 67, 75, 76, 77, 78, 79
Pinnules 59, 61, 62, 63, 64, 65, 66, 67, 68, 69, 70
Pinus (pine) 18, 44, 45, 90, 133
Pinus florissanti 90
Pinus ponderosa (yellow pine) 90
Pinus wheeleri 90
Pith 119, 120, 122, 125, 126
Pith cast 56
Pits 120, 121, 122, 123, 129, 130, 133, 134, 135
Pityocladus 40, 73, 93
Pityophyllum 40, 93

185

Pityoxylon 133
Planetree (*see Platanus*)
Platanaceae 108, 144
Platanophyllum 108
Platanus (planetree, sycamore) 42, 108, 144
Pleistocene 42, 46, 47, 48, 63
Pliocene 45, 46, 47, 89, 143, 144
Podocarpaceae 93
Podozamites 40, 94
Poison ivy (*see Rhus toxicodendron* radicans)
Pollinators 19, 25
Polypodiaceae 66
Pond scum 11
Poplar (*see Populus)*
Populus (poplar) 45, 47, 109
Precambrian 24, 35, 49
Precambrian Belt Series 49
Prickles 90
Primary veins 95
Procumbent cells 123
Prominent rays 140
Protaceae 108
Protophyllocladus 42, 93
Protopiceoxylon 40, 135
Prunus 44, 47
Pseudoctenis 40, 76
Pseudocycas 40, 78
Pseudofossils 30
Pseudotsuga (douglas fir) 44, 84
Psilophyta 12
Psilophyton wyomingensis 35
Psilotum 12
Ptelea (hop tree) 99
Pteridospermophyta (seed fern) 12, 17, 24, 68, 71
Pterocarya (wingnut) 114
Pterophyllum 40, 78
Ptilophyllum 41, 77
Puffballs 13
Pyrolusite (manganese oxide) 30

Quaternary 42, 47
Quercus (oak) 44, 45, 112, 146

Rachis 59, 60, 75, 78, 79
Radial section 117, 118
Raton Formation 23
Ray
    Biseriate 132, 133, 140, 141, 145
    Broad 146
    Heterogeneous 122, 123, 138, 139, 140, 141, 145, 147
    Homogeneous 122, 123, 138, 142, 143, 144, 147
    Multiseriate 123, 130, 138, 139, 144
    Narrow 146
    Triseriate 141, 145
    Uniseriate 122, 123, 130, 131, 132, 133, 134, 138, 139, 140, 141, 142, 143, 146, 147
Ray crossing 122, 123

Ray pitting 122
Rebuchia 35
Recent 58, 59
Red algae (*see* Rhodophyta)
Red cedar 18
Red Deer, Alberta 32
Redwood (*see Sequoia*)
Republic flora 111
Resin cells 18, 132, 134
Resin ducts (resin canals) 18, 130, 132, 133, 135
Retuse apex 98 (*see* Fig. 328)
Rhizoids 13
Rhizomes 15, 55, 58, 59, 60
Rhizopalmoxylon 148
Rhizophore 54
Rhizopus (bread mold) 12
Rhodea 37, 38, 69
Rhodophyta (red algae) 12
Rhombic-lanceolate leaf 101
Rhomboid leaf 64
Rhus (sumac) 44, 101
Rhus copallina (var. lanceolate) 101
Rhus glabra (common sumac) 101
Rhus toxicodendron radicans (poison ivy) 101
Rigbyocarpus 37, 72
Ring-porous wood 138, 143, 144, 146, 147 (Semi)
Robinia (locust) 106, 144
Rock saw 117, 118
Rockville, Oregon 32
Roots
    Fern 126
    Palm 148, 149
    Stigmaria 54
Rosa (rose) 45
Rosaceae 44, 106
Rose (*see Rosa*)
Rounded apex 70, 76, 94, 102, 104, 106, 109 (*see* Fig. 328)
Rounded base 97, 101, 104, 106, 109, 111, 112 (*see* Fig. 331)
Rubber tree (*see Hevea braziliensis*)
Ruby flora 23, 32, 45
Ruby Mountains, Montana 32
Running cypress 14
Rusts 13
Rutaceae 99, 141

Sabalites 42, 115, 116
Sac fungi 13
Saccoloma 42
Saccoloma gardneri 66
Sagebrush (*see Artemesia tridentata*)
Salicaceae 109
Salida Canyon flora 23, 36
Salina Canyon, Utah 32, 42
Salix (willow) 42, 44, 110
Salmon flora 23, 32, 44
Salmon, Idaho 32, 44
Samara 99, 103, 104

187

San Juan Basin, New Mexico 128
San Pablo flora 23, 32
Sapindus 101
Saprophytes 12
Saskatchewan
    Frenchman Formation 23, 32
Sassafras 97
Scalariform perforation plates 121, 138, 142, 144, 145
Scattered bundles 21, 147
Scheuchzer 9
Schilderia 136
Schinoxylon 141
Schinus (peppertree) 141
Schizeaceae 63
Schlotheim, von 9
Scientific name (*see* binomial nomenclature)
Scouring rushes (*see Equisetum*)
Seaweed 11
Secondary vein 95
Secondary wall 120
Seed capsule 100, 102
Seed fern (*see* Pteridospermophyta)
Seeds 37, 41, 71
    Angiosperms 20, 73, 99, 100, 103, 104, 106, 107,
    108
    Conifers 88, 90, 91, 92
    Cordaiteans 71, 72
    Cycads 73, 74
    Seed ferns 71, 72
Selaginella 14, 51
Selaginellales 14
Septate fiber tracheids 139
Septate fibers 140
Septate tracheids 134
Sequoia (redwood) 18, 41, 42, 45, 46, 88, 132
Sequoia semperviren 88
Sequoiadendron (giant redwood) 82, 84
Sequoiaoxylon 132
Serviceberry (*see Amelanchier*)
Serrate 100, 106, 107, 109, 110, 113, 114, (*see*
    Fig. 330)
Shapes 96 (*see* Fig. 329)
Shinarump Formation 130
Shoots 73, 93
Sickle shaped (*see* falcate)
Sigilaria 15, 37, 38, 51, 52, 54
Silurian 24
Simaroubaceae 99
Simple perforation plates 121, 138, 139, 140, 141,
    142, 143, 144
Sinus 96
Slime molds 13
Smithers flora 23, 40, 63
Smuts 13
Snake River 48
Snowy Range Formation 35
Solitary (isolated) vessel 138, 139, 140, 141, 142, 143,
    144, 145, 147
Sorus (fruit dot) 60, 62

188

South Africa 19
South Dakota
    Fort Union Formation 23, 32, 42, 43, 124
Spence Bridge, British Columbia 33
Spence Bridge Group 23, 33, 40
Sphagnum 13
Sphenophyllum 15
Sphenophyta (joint grass) 12, 14, 15, 16, 55
Sphenopteridaceae 69
Sphenopteridium 37, 70
Sphenopteris 37, 38, 41, 69
Spiral thickenings 142
Spokane, Washington 31
Sporangiophore 15, 56
Sporangium 14, 15, 52, 56, 62
Spore 60
Sporophyll 17, 52
Spotted Ridge flora 23, 36, 38, 55, 56, 70
Spruce (*see Picea*)
Stalk 59
Stele 120
Stem anatomy 119
Stems
    Calamites 51-52
    Conifers 129-136
    Cycads 127-128
    Dicots 137-147
    Ferns 124-126
    Lepidodendrons 55-56
    Monocots 147-148
Sterigmata 92
Sternberg 9
Stigmaria 24, 37, 54
Stinking Water flora 23, 32, 45, 99, 115
Stomata 28
Stromatolites 35, 49
Sucker Creek flora 23, 32, 45, 99, 100, 109, 114, 115
Sumac (*see Rhus*)
Sycamore (*see Platanus*)

Tangential section 117, 118
Taxales 82
Taxodiaceae 88, 132
Taxodium (bald cypress) 46, 88, 132
Taxonomy 21
Tempskya 125, 126
Tertiary 35, 42, 44, 47, 62, 63, 66, 67, 88, 89, 90, 91,
    92, 97, 101, 104, 109, 110, 113, 114, 124, 132,
    133, 141, 145
Tertiary veins 95
Teton Mountains, Wyoming 35
Tetonophycus 35
Texas 148
Thallophytes 11
Thallus 13, 49
The Dalles, Oregon 31
Thermal cement 117
Thin sections 30, 117, 118
Thouinia 44

Thorn Creek flora 23, 45, 111, 114
Thuja 86
Toothed margin (*see* dentate margin)
Torreya 94
Tracheids 18, 120, 122, 123
Transverse section 117, 118
Traumatic resin canals 129, 132
Tree ferns 59, 68
Tree of heaven (*see Ailanthus*)
Triassic 29, 38, 39, 41, 56, 58, 62, 65, 130, 131, 136
Trigonocarpus 37, 71
Triseriate rays 141, 145
Trout Creek flora 23, 31, 45, 99, 115
Truncate apex 70, 78 (*see* Fig. 328)
Truncate base 103, 104, 110 (*see* Fig. 331)
Tsuga 91
Tuber 58
Tyloses 138, 139, 140, 141, 142, 143, 144, 145, 146
Uintah flora 38
Ulmaceae 104, 143
Ulmoxylon simrothii 143
Ulmus (elm) 45, 104, 143
Umbo 90
Undulate 93 (*see* Fig. 330)
Uniseriate pitting 129, 131, 133
Uniseriate rays 122, 123, 130, 131, 132, 133, 134, 138,
    139, 140, 141, 142, 143, 146, 147
Unstoried rays (*see* nonstoried rays)
Utah
    Blackhawk flora 23
    Cedar Mountain Formation 23, 32, 139
    Chinle Formation 58, 62, 63, 64, 65, 130, 131, 136
    Dakota Formation 23, 32, 63, 64, 65, 67
    Flagstaff Limestone Formation 50
    Green River Formation 23, 32, 63, 100, 101, 102,
    105, 106, 109, 111, 114, 115
    Honaker Trail flora 23, 70
    Manning Canyon Shale 23, 33, 55, 56, 62, 68, 70,
    71, 72
    Morrison Formation 23, 33, 73, 74, 130, 134, 135
    Uintah flora 38

Valves, capsule 100
Vantage, Washington 143, 147
Vascular cylinder 120
Vascular scar 51
Vauquelinia 44
Vertical parenchyma (*see* wood parenchyma)
Vessels 120, 121, 122, 123, 137-149
Virginia City, Montana 32
Volcanism 35

Walchia 37, 84
Walnut (*see Juglans*)
Washington
    Ginkgo State Park 144
    Latah Formation 23, 31, 45, 88, 100, 114, 115
    Vantage 143, 147

Water mold 13
Wayan Formation 139
Weaverville, California 32
Weaverville flora 23, 32
Wedge shaped (*see* cuneate)
Welwitschia 19
Welwitschiales 19
Westwater, Utah 32
White 10
Whorl 55
Wieland 10
Willow (*see Salix*)
Wingatea 41, 64
Wingnut (*see Pterocarya*)
Wisconsin 48
Wood, angiosperm 137-149
Wood parenchyma 120, 121, 122, 129, 131, 132, 135,
    138, 140, 141, 142, 143, 144, 145, 146, 147
Woodworthia 39, 131
Wyoming
    Beartooth Butte flora 23
    Eden Valley   50, 140, 141, 145, 148
    Fort Union Formation 23, 32, 42, 43, 124
    Frontier Formation 67
    Green River Formation 23, 32, 63, 100, 101, 102,
    105, 106, 109, 111, 114, 115
    Medicine Bow flora 23
    Wayan Formation 139

Xenophanes 9
Xenoxylon 41, 134
Xenoxylon morrisonensis 134
Xylem 120, 121, 122

Yellow pine (*see Pinus ponderosa*)
Yellowstone 108, 144, 146
Yellowstone National Park 35, 132, 144
York Ranch flora 23

Zamia 12
Zamites 41, 77
Zeilleria 70
Zelkova 43, 45, 115
Zygopteridaceae 62

Leaf Characters

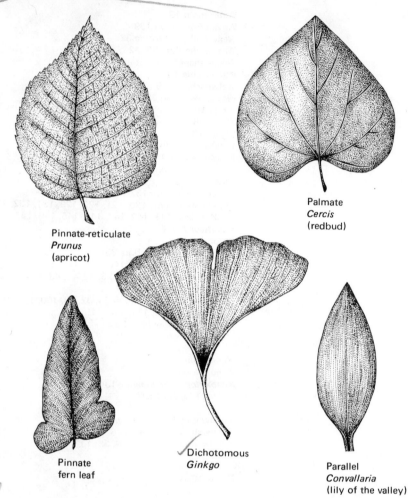

Pinnate-reticulate
Prunus
(apricot)

Palmate
Cercis
(redbud)

Pinnate
fern leaf

Dichotomous
Ginkgo

Parallel
Convallaria
(lily of the valley)

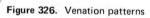

Figure 326. Venation patterns

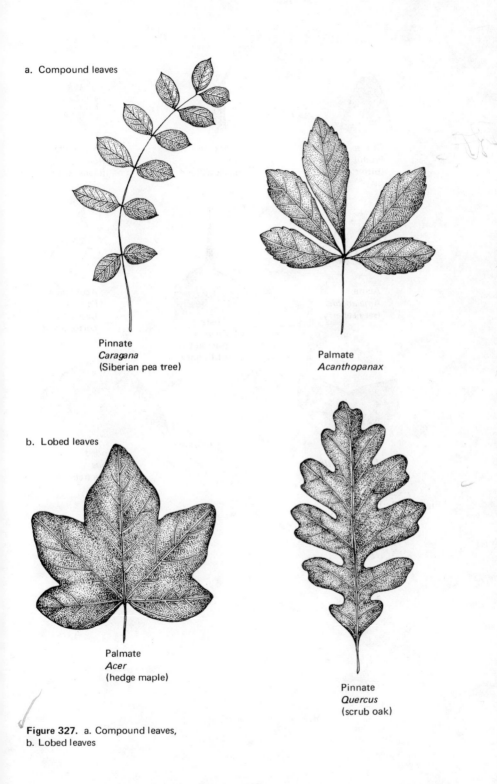

a. Compound leaves

Pinnate
Caragana
(Siberian pea tree)

Palmate
Acanthopanax

b. Lobed leaves

Palmate
Acer
(hedge maple)

Pinnate
Quercus
(scrub oak)

Figure 327. a. Compound leaves,
b. Lobed leaves

Obtuse
Prunus
(bitter cherry)

Attenuate
Salix
(black willow)

Emarginate
Robinia
(black locust)

Acute
Amelanchier
(serviceberry)

Aristate
Ficus
(sacred tree
of India)

Acuminate
Populus
(black
cottonwood)

Truncate
Liriodendron
(tulip tree)

Rounded
Cotinus
(smoke tree)

Figure 328. Leaf apices

Orbicular
Symphoricarpos
(snowberry)

Elliptic
Quercus
(laurel oak)

Oblong
Rhododendron

Linear
Salix
(desert willow)

Lanceolate
Salix
(black willow)

Oblanceolate
Bumelia
(gum bumelia)

Ovate
Fraxinus
(single-leaf ash)

Oval
Cornus
(dogwood)

Cordate
Syringa
(lilac)

Obovate
Nyssa
(black gum)

Deltoid
Populus
(eastern cottonwood)

Figure 329. Leaf shapes

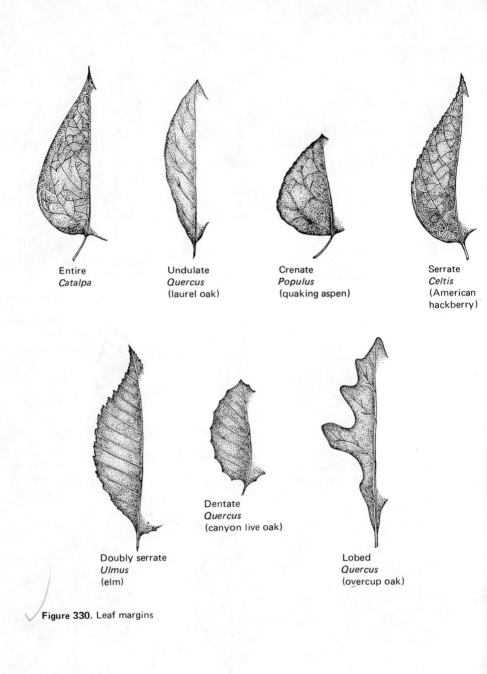

Entire
Catalpa

Undulate
Quercus
(laurel oak)

Crenate
Populus
(quaking aspen)

Serrate
Celtis
(American
hackberry)

Doubly serrate
Ulmus
(elm)

Dentate
Quercus
(canyon live oak)

Lobed
Quercus
(overcup oak)

Figure 330. Leaf margins

Acute
Cornus
(dogwood)

Cuneate
Magnolia

Attenuate
Quercus
(white oak)

Asymmetrical
Ulmus
(elm)

Rounded
Betula
(yellow birch)

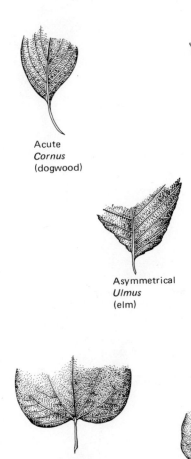

Cordate
Cercis
(redbud)

Auriculate
Camptosorus
(walking fern)

Obtuse
Alnus
(red alder)

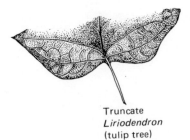

Truncate
Liriodendron
(tulip tree)

Figure 331. Leaf bases

Completed
23 July 1976